MARKUS VARESVUO & MIKKO OIVUKKA

BIRDS OF FINLAND

A PHOTOGRAPHIC GUIDE

H E L M

LONDON • OXFORD • NEW YORK • NEW DELHI • SYDNEY

HELM
Bloomsbury Publishing Plc
50 Bedford Square, London, WC1B 3DP, UK
Bloomsbury Publishing Ireland Limited,
29 Earlsfort Terrace, Dublin 2, D02 AY28, Ireland

First published in the United Kingdom 2026

A catalogue record for this book is available from the British Library.
Library of Congress Cataloguing-in-Publication data has been applied for.

ISBN: PB: 978-1-3994-2396-0;
ePub: 978-1-3994-2397-7; ePDF: 978-1-3994-2394-6

2 4 6 8 10 9 7 5 3 1

Design by Susan McIntyre
Map by Julian Baker

Printed and bound in Slovenia by DZS Grafik

CONTENTS

INTRODUCTION

Finland lies in eastern Scandinavia, between Sweden and Russia, making it a dream destination for birdwatchers across Europe. It is a land where northern, eastern and southern species meet, famous for its vast wilderness where people still live in harmony with bears and wolves. While it is known as the 'land of a thousand lakes', Finland also depends heavily on its forest industry, which results in lots of species-poor coniferous forest and logging. Combined with the Nordic climate, this means bird numbers can be limited, and species are sometimes difficult to find without knowing the right locations.

Finland is the fourth-largest country located entirely in Europe, offering a wide range of habitats. The terrain is mostly flat, though mountains along the Norwegian border rise to about 1,300m. The Baltic Sea coastline, stretching from the Russian border north to the Oulu region, together with thousands of islands in numerous archipelagos, provides excellent areas for seabirds and stopovers for migrating waders and warblers. Many raptors also follow the coastline, avoiding the long journey across open sea. Inland lakes serve as important staging sites for waterfowl during migration, while the eastern border forests hold fascinating local specialities. Lapland is almost entirely tundra, favoured by northern species.

Understanding the influence of climate on birdlife is important before travelling to Finland. Every season has its own interesting features, and Finnish nature offers

Black Grouse.

an unforgettable experience all year round. Winters can be harsh, especially in the north, and only a few species manage to survive through this season. Days are very dark, with only a few hours of daylight for birding. In contrast, spring is the golden time for birdwatchers: nature awakens, and migratory birds rush north to their breeding grounds. Summer is the season for breeding birds, when birdwatchers need to observe carefully to avoid disturbing them. In midsummer, birding often takes place when it is nighttime at lower latitudes, since the sun remains above the horizon. In autumn, species gradually disappear unnoticed one by one, and only a few show more noticeable visible migration.

While local birders focus on spring and autumn migration, always hoping for rare vagrants, birdwatching visitors come mainly for three reasons: extraordinary residents such as owls and grouse, eastern specialities like bluetails and buntings, and special northern birds. Kuusamo is the number one destination for bird tourists, because this is where almost all of Finland's most exciting species can be found.

This book is designed for people planning to visit Finland to enjoy nature. It presents 300 species, including all the regular breeding birds and the most common vagrants. With a combination of beautiful photographs and short descriptions, it offers an easy-to-carry guide for use in the field.

SEASONS

Finland is a country of four distinct seasons, each offering unique opportunities for birdwatchers. If you plan to visit in search of certain species, it is important to know when they can be found. Conditions also vary greatly between the south and north: in the north snow often melts only in May, while in the south it may already feel like summer.

WINTER
Winter in Finland is harsh, and only a few resident species remain through the season. Snow conditions vary from year to year: in the south snow cover can be thin or absent, while in the north it usually remains deep until spring. Bird feeders are very common and provide an important supplement that helps species survive. If the Baltic Sea stays open, some waterbirds may overwinter, but if it freezes completely, almost all migrate further south.

SPRING
Spring is the golden season for birdwatchers. Migratory birds from Europe and Africa first arrive in southern Finland before gradually continuing north to their breeding grounds. The migration begins in March and lasts until the end of May. Monitoring is especially active on islands and along the coasts, with the highlight being the spectacular passage of Arctic-breeding waterfowl and waders along the eastern Gulf of Finland.

Male Ruffs fighting.

Spring is also the main season for courtship displays. Black Grouse, Capercaillie, Ruff and many other species are especially active, and all birds appear in their most striking breeding plumage. Singing and display behaviour make them easier to observe, offering some of the best birdwatching of the year.

SUMMER

Summer is a season of nightless 'nights', when the sun barely sets in the north. Birdwatching is often most rewarding during the night hours, as many species are then active and light conditions are ideal for photography.

All birds nest during summer, and it is important to avoid disturbance to ensure that they can breed successfully. Most species are incubating in May, and chicks hatch in June, though timing differs between south and north. Some species raise multiple broods, remaining in the breeding cycle throughout the summer.

By midsummer, signs of autumn migration appear. Waders begin their southward journey in July and August and can be seen feeding along shores.

AUTUMN

Autumn is the season when many birds leave Finland for their wintering areas. While some depart in July and August, others remain until late October or November. Cranes, geese and swans migrate in large flocks, often gathering in fields in groups of thousands before departure. Other species disappear gradually and less noticeably, often migrating at night. This makes autumn migration both fascinating and elusive for birdwatchers.

HABITATS

Finland offers a wide range of habitats, from forests and wetlands to coasts and farmland. Each provides unique conditions for birds, and together they support a rich diversity of species throughout the year.

ISLANDS

The many islands and archipelagos are important bird areas, especially during migration and the breeding season. Seabirds, gulls, terns and waders are typical, and some islands host large colonies of nesting birds. Many small birds also stop at the islands during migration, making these areas important resting and feeding sites.

LAKES AND RIVERS

Thousands of lakes and countless rivers make freshwater habitats dominant in Finland's landscape. These waters are home to swans, divers, grebes, ducks and other waterbirds. Shoreline vegetation also provides shelter for warblers and small passerines. Some species prefer freshwater lakes, while others are found in lush or swampy lakes.

COASTAL WETLANDS AND BAYS

Shallow bays and coastal wetlands are among the richest bird habitats in Finland. They are especially important during spring and autumn migration, attracting large numbers of waterfowl and waders.

FIELDS

Open farmland is an essential feeding and resting ground, particularly during migration. Cranes, geese and swans gather in large flocks, while Skylarks and wagtails are often seen in summer.

SWAMPS AND BOGS

Vast peatlands are characteristic of Finland, especially in the north. They support species such as Willow Grouse, Golden Plover, Jack Snipe

A flock of Willow Grouse.

and Cranes. These wetlands are also important breeding areas for many insect-eating birds.

CONIFEROUS FORESTS
Spruce and pine forests cover much of the country. They provide habitats for Capercaillie, crossbills, woodpeckers, owls and many forest passerines.

MIXED AND DECIDUOUS FOREST
Birch, aspen, alder and mixed woodlands create more diverse conditions than coniferous forests. Typical species include flycatchers, tits, thrushes and several migratory songbirds.

BEACHES AND SHORES
Sandy, rocky and wetland shores are valuable stopover sites for waders, gulls and terns during migration. Some species, such as Ringed Plover, also breed along the coast.

URBAN ENVIRONMENTS
Towns and villages attract adaptable birds such as House Sparrow, Jackdaw, gulls and pigeons. Parks and gardens provide habitats for Blackbird, tits and woodpeckers, showing how wildlife can thrive in human settlements.

MEADOWS AND PASTURES
Traditional meadows and pastures, though less common today, are important for open-land birds such as Skylark and Whinchat. They also serve as feeding areas during migration.

MOUNTAINS AND TUNDRA
Cliffs and rocky hillsides are key nesting sites for raptors such as Eagle Owl, Peregrine Falcon and Gyrfalcon, as well as Raven. Mountain tundra in Lapland supports specialised species like Rock Ptarmigan, Dotterel, Long-tailed Skua and various waders. Despite harsh weather and a short growing season, tundra remains an important breeding ground for Arctic birds and offers rare opportunities to observe species adapted to extreme environments.

Arctic Tern.

HOW TO FIND OWLS IN FINLAND

Finland is well known for its owls, and for many birdwatchers seeing the magnificent Great Grey Owl is a lifelong dream. However, visitors often assume that spotting one is easy. The truth is quite the opposite. Owls are unpredictable and secretive, and seeing them in daylight requires patience and a bit of luck. Almost all species are nocturnal, active mainly at night or during the dim light of dawn and dusk.

WHERE AND WHEN TO LOOK

Finland is one of the best countries in Europe to observe owls. With nine regularly occurring species and vast areas of forest, wetland and open countryside, it offers superb opportunities for both beginners and experienced birders. The best time to search for owls is from March to June, when most species are breeding. Males begin to call in late winter to establish territories, and their vocal activity peaks in early spring. Calm, clear evenings in March and April are perfect for listening. Even a short stop along a quiet forest road might reward you with the calls of Tawny, Tengmalm's, Ural or Pygmy Owls.

HABITATS AND SPECIES

Each owl prefers its own type of habitat. Tawny Owl is the most common in southern Finland, nesting in mixed forests, parks and even near villages. Ural Owls inhabit mature pine and spruce forests, especially in central and eastern parts of the country. The tiny Pygmy Owl lives in coniferous forests and is best detected by its whistled call at dawn or dusk. In Lapland, you might encounter Hawk Owl, which hunts from treetops or poles in open forest areas, or the ghostly Great Grey Owl, which favours large bogs and forest edges.

HOW TO FIND THEM

Listening for calls on calm nights is one of the most effective ways to locate owls. Each species has its own distinctive voice from the deep hoots of Tawny Owl to the rhythmic 'pu-pu-pu' of Tengmalm's Owl. During the day, look for revealing signs such as white droppings, pellets or feathers under roosting trees.

Many owls, including Tengmalm's, Ural and Pygmy, use nest boxes. Networks of boxes maintained by volunteers, bird clubs and local landowners often provide good viewing opportunities in spring. Hiring an experienced birdwatching guide is the most reliable way to find owls, especially around Oulu and Kuusamo, the best regions for owl watching in Finland.

A MEMORABLE ENCOUNTER

In winter, Hawk Owls and Great Grey Owls sometimes move south in search of food, appearing surprisingly close to towns and roads. During such years, they can even be seen perched on roadside poles or hunting over fields.

Finding an owl in Finland is always a magical experience. Whether you hear the haunting call of a distant Ural Owl or watch a silent Great Grey glide across a snowy bog at dusk, the encounter will leave you with an unforgettable memory of Finland's wild and mysterious nature.

BIRDWATCHING IN FINLAND

Birdwatching is very popular in Finland. Many Finns are interested in birds, and a large number are members of the Finnish birdwatching association, BirdLife Finland, which is part of BirdLife International. BirdLife Finland has around 30 local associations that work voluntarily to help protect birds. Their activities include bird counting, maintaining and renovating birdwatching sites and organising birdwatching walks for beginners.

In Finland, only some birders use eBird. Instead, BirdLife Finland manages its own system called Tiira. In the Tiira database, anyone can submit observations, but only members of BirdLife Finland can access and read them.

Birdwatching is generally easy in Finland. Travelling by car is convenient – main roads are in good condition, traffic is calm and routes are clearly signposted. Most of the best birdwatching sites have observation towers, providing easy access and good watching points for birdwatchers. However, birdwatching in the archipelagos can be more challenging, as some locations require boat transportation.

Hunting is a common activity in Finland during autumn. For this reason, many game species such as grouse are very shy and tend to disappear quickly when disturbed. Illegal hunting is rare, and most hunters are responsible, hunting only within regulations and for their own use. For birdwatchers and photographers, a car often works as an excellent hide, allowing close observation without alarming the birds.

Hawk Owl.

Finland is full of wonderful birdwatching sites. Most are located along the coast, but the inland areas also hold many hidden gems. In the section 'Best birdwatching sites in Finland' we highlight locations that are easy to reach and well equipped for visitors. Most sites offer good parking, and many have birdwatching towers for excellent views. On some islands, accommodation and transport services are also available.

There are not many professional bird guides in Finland, as most birders are hobbyists and work voluntarily. However, in regions such as Oulu and Kuusamo, a few experienced guides are available to help visitors find their target species.

BEST BIRDWATCHING SITES IN FINLAND

Finland is a land of quiet beauty, where endless forests, peaceful coasts and countless islands create perfect homes for birds. Among the archipelagos and along the shorelines, nature feels alive, and in Lapland the stillness of the fells offers something truly special.

For anyone who loves exploring, knowing the best birdwatching spots makes the experience even more rewarding – places where nature feels close and every sighting becomes a small treasure. In this guide, we've chosen sites that offer a good chance to see many of Finland's most interesting bird species and where the infrastructure is well organised. Of course, there are many other wonderful locations to discover throughout the country.

AHVENANMAA (ÅLAND)

Åland, the largest island of the Åland Archipelago in the northern Baltic Sea between Sweden and Finland, offers a wide variety of habitats for birdwatchers. Coastal shores, rocky skerries, reedbeds, pastures and forests support rich birdlife throughout the year.

Spring and autumn bring spectacular migrations, as thousands of geese, ducks and waders pass through. Many passerines stop on the islands on their way to breeding grounds, providing great opportunities to find interesting species in the bushes and along field edges.

Seabird colonies, including auks, gulls and terns add to the appeal, while raptors such as White-tailed Eagles often patrol the skies.

The main island is easily accessible by road, while smaller islands require boat transport to reach the best birding hotspots. You can reach Åland by ship or ferry from Finland or Sweden.

KÖKAR ISLAND

Kökar is a remote gem in the Åland Archipelago, prized for its rich mix of coastal and island birdlife. Windswept shores, sheltered bays, rocky cliffs and meadows create a variety of habitats for both breeding and migrating birds.

During spring and autumn migration, large flocks of geese, divers and ducks pass through – including Common Eider, scoters, Long-tailed Duck, Barnacle and Greylag Geese and Black-throated Diver, among many others. The island's lagoons gather impressive numbers of ducks such as Tufted Duck, Scaup, Wigeon and Shoveler.

Bushes, fields and small forests shelter a wide range of passerines – shrikes, chats and warblers are all possible. Kökar is also the best place in Finland to find Barred Warbler during the summer breeding season.

Kökar's charm, peaceful atmosphere and rich birdlife make it a rewarding destination for dedicated birders. The island can be reached by ferry from mainland Finland or from Åland.

Utö's eastern meadow.

ÖRÖ ISLAND
One of the largest islands in the Archipelago Sea, Örö offers remarkably diverse habitats: rocky shores, sheltered bays, coastal meadows, juniper-covered grasslands, forests and historic landscapes shaped by its military past.

During spring and autumn migration, flocks of ducks and skuas pass offshore, including scoters, Long-tailed Ducks and Arctic Skuas. The shores host waders such as Greenshank, and Common and Spotted Redshanks, while Barnacle Geese and Common Eiders are regular along the coastline. Coastal meadows come alive in summer with Red-backed Shrikes, Whinchats, pipits and Skylarks. In autumn Örö is a good place for eastern vagrants such as Yellow-browed and Pallas's Leaf Warblers.

Raptor migration can also be impressive on Örö, with hundreds of Sparrowhawks, harriers and falcons passing through on good days.

Örö is accessible by ferry or private taxi boat, with regular summer connections, and can also be reached year-round – offering rewarding birdwatching in every season.

UTÖ ISLAND
Finland's southernmost island, Utö, is famed for its windswept shores, rocky cliffs and coastal meadows – a true outpost for bird migration in the Baltic Sea.

As a key stopover point, Utö hosts large flocks of waders such as Purple and Curlew Sandpipers, while Arctic Skuas and both Arctic and Common Terns patrol the surrounding waters. The island's reedbeds, meadows and small patches of scrub attract a wide range of passerines, including Bluethroat, Semi-collared Flycatcher and other sought-after migrants. During spring and autumn, Utö is one of Finland's best places to observe rare vagrants and the thrill of migration days when new species appear overnight. In autumn, Yellow-browed and Pallas's Leaf Warblers are regular highlights.

Compact and easy to explore on foot, Utö is accessible year-round by ferry from the mainland and offers comfortable accommodation and restaurants – an ideal destination for birders seeking both ease and excitement.

JURMO ISLAND

Neighbouring Utö, Jurmo is a completely different kind of island – long, low and windswept, with rocky beaches, reedbeds and coastal meadows stretching towards the open sea.

During migration, the island comes alive with flocks of waders such as Dunlin, Ruff and Bar-tailed Godwit, alongside sandpipers including Curlew Sandpiper, Sanderling, Red Knot and Little Stint. The island's small forest patches and bushes also shelter a fine variety of passerines and often reveal rare vagrants, making Jurmo an exciting place for dedicated birders.

In summer, Little Terns may breed on the shores, and Sandwich Terns are occasionally seen fishing offshore. Also, Jurmo is the best place to see Purple Sandpiper in winter.

Jurmo's length and rocky terrain make birding here more demanding than on neighbouring Utö, but its wild beauty and richness of birdlife reward every visitor. Parts of the island are closed during the breeding season to protect nesting birds, ensuring peace for both wildlife and watchers.

ISOKARI ISLAND

Isokari, a small yet iconic island crowned by its historic lighthouse, offers a rich mix of rocky shores, coastal meadows, forest patches and open waters.

The harbour is home to a colony of Black Guillemots, while Razorbills can often be seen nearby during summer boat trips. Spring and autumn migration bring a steady passage of waders and ducks along the coast, while the meadows host Whinchat, Northern Wheatear and Red-backed Shrike. In autumn, rarities such as Yellow-browed and Pallas's Leaf Warblers may also appear among the bushes.

During summer, Common and Arctic Terns nest in lively colonies, their calls filling the air, and raptors like White-tailed Eagle, Eurasian Sparrowhawk and Marsh Harrier frequently patrol overhead.

Accessible by boat, Isokari combines remote island charm, striking coastal scenery and rewarding birdwatching opportunities – especially during the vibrant migration seasons.

RUISSALO ISLAND

Just outside Turku, Ruissalo Island is one of southwest Finland's best-known birdwatching destinations. The island's rich oak forests, meadows and scenic shoreline create diverse habitats that attract both forest and coastal species.

In spring, Ruissalo comes alive with birdsong. The ancient oak woods around Marjaniemi and the Botanical Garden are excellent for forest birds, including Tawny and Pygmy Owls, Grey-headed and Lesser Spotted Woodpeckers, Hawfinch, Common Rosefinch, and Icterine and Wood Warblers.

Kansanpuisto's coastal meadows and bays are good spots for waders and ducks, while Kuuva and Saaronniemi provide fine views of migrating waterbirds and gulls. In autumn, thrushes, tits and finches gather in large flocks before heading south, and White-tailed Eagles can often be seen patrolling offshore.

With its combination of rich birdlife, easy access by car, bicycle or public transportation, and beautiful landscapes,

Ruissalo offers one of the most enjoyable birdwatching experiences on Finland's southern coast.

MIETOISTENLAHTI (MIETOINEN BAY)

Located on Finland's southwest coast, Mietoistenlahti is a rich and varied birdwatching destination. Its mix of sheltered bays, reedbeds, mudflats, fields, lagoons and estuarine shores supports a wide diversity of species year-round.

During spring and autumn migration, the bay comes alive with ducks and waders such as Wigeon, Teal, Redshank, Dunlin, Golden Plover and a variety of gulls and terns, making it an exciting stopover for migrating birds. Thousands of geese, including Taiga and Tundra Bean Geese and Greater White-fronted Goose, and flocks of Cranes often stop to feed in nearby fields. Raptors, including Peregrine Falcon, White-tailed Eagle and harriers, patrol the skies, adding drama as they hunt over the bay.

In summer, reedbeds and coastal meadows host Reed and Sedge Warblers, Common Rosefinch, Whinchat, Red-backed Shrike, Yellow Wagtail and many other species. With accessible bird towers and vantage points, Mietoinen Bay offers both easy access and rewarding birdwatching throughout the year.

HALIKONLAHTI (HALIKKO BAY)

Halikonlahti, near Salo, is one of Finland's top birdwatching spots, offering a rich mix of reedbeds, mudflats, ponds and coastal meadows that serve as vital feeding and resting grounds for birds all year.

During spring and autumn migration, the bay comes alive with geese and ducks such as Pochard, Gadwall, Taiga Bean and Tundra Bean Geese and Whooper Swan. Mudflats attract waders including Ruff and Common Redshank, while raptors such as White-tailed Eagle and Osprey patrol overhead. Migrating passerines often stop here, adding to the diversity.

In summer, Halikonlahti is home to breeding species such as Marsh Harrier, Thrush Nightingale, and Reed and Sedge Warblers. Reedbeds shelter Reed Bunting and other songbirds, while open waters host Great Crested Grebe and Goosander.

With its well-equipped bird observatory, hides and observation towers, Halikonlahti offers both easy access and outstanding birdwatching throughout the seasons.

KIRKKONUMMI PORKKALA

The Porkkala Peninsula, southwest of Helsinki in Kirkkonummi, is one of southern Finland's finest seawatching and migration sites. Extending far into the Gulf of Finland, it provides superb vantage points for observing both seabirds and raptors on the move.

In spring, vast numbers of Arctic waterfowl including Long-tailed Duck, scoters, eiders, divers and geese pass close to shore on their journey eastward. Autumn brings impressive movements of divers, skuas and other seabirds heading south. White-tailed Eagles are resident year-round, and are often seen gliding along the coast, while the peninsula's forests, rocky shores and small bays host a wide variety of breeding birds. Fields along the approach roads are also good for migrating Cranes and geese during spring and autumn.

Easily reached from the Helsinki region, Porkkala combines dramatic coastal scenery with rich birdlife, making it one of Finland's most rewarding and accessible birdwatching destinations.

HANKO TULLINIEMI (UDDSKATAN)

Tulliniemi, the southernmost tip of mainland Finland, is a celebrated birdwatching site where coastal cliffs, sandy shores and open meadows meet. Its position at the entrance to the Gulf of Finland makes it a key stopover for migratory birds, especially during spring and autumn.

Spring migration brings a rich variety of passerines and seabirds. Whinchats and Northern Wheatears frequent the rocky shorelines and meadows, while coastal waters host flocks of Common Eider and Velvet Scoter. Along the coast, Black-headed and Herring Gulls are common, and Sandwich Terns are occasionally seen offshore.

Sheltered bays and reedbeds attract numerous warblers, while surrounding pine groves and hedgerows hold species such as Red-backed Shrike and Tree Pipit. Raptors, including White-tailed Eagle and Peregrine Falcon, are often seen, as many other migrants pass through the peninsula.

Accessible on foot from Hanko, Tulliniemi offers marked trails, scenic viewpoints and the Halias Bird Observatory – a hub for birders during migration periods.

ESPOO SUOMENOJA (FINNOO)

Suomenoja, also known as the Finnoo wetlands, is one of the best-known urban birdwatching sites in southern Finland. This small but species-rich wetland is famous for its large breeding colony of Black-headed Gulls, which provides protection for many other birds nesting in the area.

The ponds and reedbeds attract numerous waterbirds, including Great Crested and Little Grebes, rails and crakes, Pochard, Gadwall and Shoveler. Reed and Sedge Warblers, and other passerines are common in summer, while Marsh Harriers and Hobbies are occasionally seen hunting over the area. During migration, the wetlands also serve as a stopover site for waders and ducks.

Suomenoja is easily accessible by public transport and has well-maintained paths, making it a convenient birding location close to Helsinki and Espoo. Its combination of rich birdlife and urban surroundings makes it a favourite destination for both beginners and experienced birdwatchers.

NUUKSION KANSALLISPUISTO (NUUKSIO NATIONAL PARK)

Nuuksio National Park, located just outside Helsinki and Espoo, offers a completely different birdwatching experience from Finland's coastal and wetland sites. Its landscape of deep forests, clear lakes and rocky ridges provides habitats for a wide variety of forest species.

Typical birds include Capercaillie, Hazel Grouse and several woodpecker species, including Grey-headed and Three-toed. The quiet lakes sometimes host Red-throated Divers, while the surrounding woods are home to forest owls – Ural, Pygmy and Tengmalm's Owls all breed here. In summer, the park comes alive with the songs of Red-breasted Flycatcher, Wood Warbler and other passerines.

Easily accessible from the capital region, Nuuksio offers marked trails, observation points and good facilities. For birdwatchers, it provides an ideal opportunity to experience Finland's rich forest birdlife in a peaceful, natural setting – just a short journey from the city.

ESPOO LAAJALAHTI

Laajalahti Nature Reserve, located between Espoo and Helsinki, is one of southern Finland's most important wetland bird areas. Its shallow bay, wide reedbeds and coastal meadows form excellent habitats for both breeding and migrating birds throughout the year.

In spring and autumn, Laajalahti becomes a vital stopover for waterfowl and waders. Breeding species include Marsh Harrier, Eurasian Bittern and various ducks and warblers. The reedbeds are home to Great Reed Warbler and Bearded Reedling, while White-tailed Eagles are often seen soaring above the bay. Great Egrets also visit regularly, adding to the excitement.

The reserve features several bird-watching towers, well-marked trails and a nature information centre, offering excellent access and facilities for visitors. Combining rich birdlife, scenic wetlands and easy reach from the capital region, Laajalahti is a must-visit destination for birdwatchers exploring the Helsinki area.

HELSINKI VIIKKI AND VANHANKAUPUNGINLAHTI

Located in northeastern Helsinki, just a few kilometres from the city centre, Viikki and Vanhankaupunginlahti together form one of Finland's best-known urban birdwatching areas. This nature reserve offers a rare combination of wetlands, coastal shores, open meadows and small patches of forest – a peaceful haven for both breeding and migratory birds right next to the capital.

The wide reedbeds and mudflats of Vanhankaupunginlahti attract an impressive variety of waders and waterbirds during migration seasons. Curlew Sandpiper, Red Knot, Sanderling and Dunlin can be

Barnacle Geese at the Helsinki Viikki.

seen feeding along the shores, while Common and Arctic Terns, along with Black-headed Gulls, patrol the bay. In spring and autumn, the skies are alive with movement, and on calm days, the sounds of calling waders fill the air.

At Viikki, reedbeds and meadows support breeding Reed and Sedge Warblers, Bearded Reedling and Thrush Nightingale. Northern Wheatear, Whinchat and Yellow Wagtail are common during migration, while wooded edges often conceal Red-backed Shrike, Tree Pipit and Common Rosefinch. Overhead, Marsh Harrier, Osprey and White-tailed Eagle are frequently seen hunting. Thousands of Barnacle Geese live on the fields and often fly overhead.

Easily accessible by public transport, Viikki offers well-marked trails, bird towers and hides – a perfect escape for anyone wishing to experience rich birdlife within the heart of Helsinki.

VIROLAHTI HURPPU

Virolahti, on Finland's southeastern coast near the Russian border, is one of the country's best-known birdwatching areas, particularly famous for its spectacular spring migration. The coastal waters, bays and islands around Hurppu form a vital migration corridor for Arctic waterfowl travelling along the Gulf of Finland.

In May, thousands of Long-tailed Ducks, Common and Velvet Scoters, and various geese and divers pass close to shore, often in vast, swirling flocks when winds are favourable. Waders, gulls and terns follow the same route, while raptors such as White-tailed Eagle and Rough-legged Buzzard are frequently seen patrolling the coast. The Virolahti area is also one of the best places in Finland to spot eastern raptors such as Greater Spotted Eagle and Black Kite arriving from the Russian side.

Observation towers and well-placed viewpoints around Hurppu provide excellent opportunities to witness this impressive migration spectacle.

PARIKKALA SIIKALAHTI

Siikalahti in Parikkala is a nationally recognised Important Bird and Biodiversity Area (IBA), renowned for its rich variety of habitats, including a shallow lake, reedbeds, meadows and surrounding forests. These diverse environments support exceptional birdlife throughout the year.

Spring and early summer bring large numbers of breeding wetland birds such as crakes, bitterns, marsh harriers, ducks and waders. Cranes and Whooper Swans breed regularly, while reedbeds resound with the songs of Reed and Blyth's Reed Warblers. Occasional highlights include Booted Warbler, Golden Oriole and White-backed Woodpecker. Migration seasons bring spectacular movements of thousands of geese, swans and ducks, together with waders and raptors.

Well-maintained trails and observation towers make Siikalahti both accessible and rewarding for visitors, while careful protection ensures its wildlife thrives. This combination of habitat richness and accessibility makes Siikalahti one of Finland's finest and most celebrated birdwatching destinations.

PORI YYTERI

Yyteri, on Finland's west coast, by the Bothnian Sea, is renowned for its vast

Barnacle Geese during arctic migration in Virolahti.

sandy beaches, dune forests, reedbeds and wetlands. This variety of habitats makes it a superb destination for birdwatching year-round, especially during migration seasons.

Spring and autumn bring impressive movements of shorebirds and gulls feeding along the tidal flats. Red Knot, Sanderling, Curlew Sandpiper and various sandpipers are among the highlights, while large flocks of gulls patrol the coast. Common Shelducks form large flocks on the beach. Dune forests and reedbeds provide shelter for passerines including Whinchat, Northern Wheatear, and Reed and Sedge Warblers. Coastal meadows attract Skylark and Meadow Pipit, while raptors such as Marsh Harrier and White-tailed Eagle can be seen. In summer, Yyteri hosts breeding colonies of Common and Arctic Terns.

Marked trails, observation towers and visitor facilities make Yyteri both accessible and rewarding, offering an exceptional birdwatching experience where dramatic coastal scenery meets vibrant avian life.

PORI ENÄJÄRVI

Enäjärvi, a freshwater lake near Pori, is surrounded by reedbeds, marshes and wet meadows, creating rich habitats for a wide range of bird species throughout the year. Its calm waters and sheltered shores make it an important site for both resident and migratory birds.

Spring and autumn migrations bring flocks of dabbling ducks such as Teal and Gadwall, while waders like Wood and Green Sandpipers feed along the shorelines. Reedbeds host singing Reed and Sedge Warblers, and breeding species include Reed Bunting and Water Rail. Open waters attract Great Crested Grebe, Tufted Duck and occasional diving ducks. Raptors such as Hobby and Marsh Harrier are often seen overhead.

Well-maintained trails and bird towers make Enäjärvi highly accessible, offering birdwatchers the chance to enjoy peaceful observation in a tranquil wetland setting rich in avian diversity.

PORI LEVEÄKARI

Leveäkari, located on the coast of Pori near Yyteri, is one of the finest birdwatching sites along the Bothnian Sea. This small coastal area combines mudflats, rocky shores, islets and shallow bays, creating ideal habitats for waders, gulls and waterfowl. During spring and autumn migration, Leveäkari attracts large flocks of Dunlin, Ringed Plover, Common Redshank and Curlew, often joined by scarcer species such as Citrine Wagtail, Great Egret, Curlew Sandpiper, Little Stint and Broad-billed Sandpiper.

In summer, the area supports breeding Arctic and Common Terns, while the surrounding waters host various duck species, including Goosander and Goldeneye. Offshore rocks are favoured perches for Cormorants and White-tailed Eagles, the latter often seen gliding over the bay. During calm migration days, migrating passerines and raptors can also be observed moving along the coast.

Leveäkari is easily accessible by car and offers wide coastal views, making it an excellent spot for both casual birders and keen birdwatchers visiting the Pori region. Note: the coastal meadow is closed in summer to protect breeding birds.

PORI REPOSAARI

Reposaari, located off the coast of Pori at the mouth of the Kokemäenjoki River, is a unique and rewarding destination for birdwatchers. This small island, connected to the mainland by a causeway, offers a rich variety of habitats from coastal meadows and rocky shores to lush gardens and mixed forests, making it attractive to both breeding and migratory birds.

In spring and autumn, Reposaari becomes a migration hotspot where birders can witness impressive movements of passerines, thrushes and raptors arriving from or departing across the sea. The surrounding waters and shorelines host flocks of Long-tailed Duck, Goldeneye, Goosander and various gulls.

In autumn, Reposaari is especially known for its eastern vagrants with Yellow-browed, Pallas's Leaf and Dusky Warblers among the regular highlights.

Accessible trails, lookout points and proximity to Yyteri make Reposaari an ideal complement for birdwatching in the Pori region where sea, forest and migration routes beautifully converge.

KRISTIINANKAUPUNKI SIIPYY

Siipyy, near Kristiinankaupunki on the Bothnian Sea coast, offers a rich and varied coastal landscape of sandy shores, river mouths, reedbeds and meadows. Its strategic location along the migration route makes it one of the best places in western Finland to witness the spectacle of migration, with thousands of birds passing overhead each spring and autumn.

The area is excellent for seabirds, including Common Guillemots, Razorbills, Arctic Skuas, Caspian Terns, Black and Velvet Scoters, Long-tailed Ducks, Common Eiders, grebes, and Black-throated Divers, with the occasional Sandwich Tern also seen.

During migration, large flocks of waders, including Dunlin, Ruff, Red Knot, Curlew and Bar-tailed Godwit fly along the coastline in impressive numbers. Reedbeds and

Whoopers Swans during spring at Liminka Bay.

wet meadows are alive with Sedge and Reed Warblers, Bluethroats, and Whinchats, while Red-backed Shrikes, Common Rosefinches and Yellow Wagtails frequent nearby fields and bushes. Overhead, White-tailed Eagles and migrating Ospreys are regularly seen.

In summer, Common and Arctic Terns breed along the coast, joined by Oystercatchers and Lapwings. With its well-marked trails and bird towers, Siipyy is one of Finland's most rewarding coastal birdwatching destinations.

HAILUOTO

Hailuoto, the largest island in the northern Gulf of Bothnia, lies just off Oulu and offers an outstanding variety of habitats – sandy beaches, dunes, coastal meadows, forests and wetlands. Its strategic location makes it one of Finland's most important migration stopovers, attracting huge numbers of geese, swans, Cranes, ducks, waders, gulls and terns during spring and autumn. Passerines also gather here in large numbers, and rarities are regularly recorded, especially during migration peaks.

In summer, Arctic Terns and several wader species breed along the shores. Common Shelduck is regularly seen along the coast, while forests and meadows echo with woodpeckers, crossbills and many other species. Observation towers and marked trails make exploring easy, offering superb vantage points. Hailuoto's mix of habitats and spectacular migration ensures there is always something to see for birdwatchers.

LIMINGANLAHTI (LIMINKA BAY)

Liminka Bay, on the northern shore of the Gulf of Bothnia near Oulu, is one of Finland's premier birdwatching sites. This expansive wetland of reedbeds, meadows and mudflats provides vital habitat for thousands of waterbirds, waders and passage migrants.

Spring and autumn are especially spectacular, as the bay serves as a key stopover along major migratory routes. Large flocks of Cranes, Greylag

Siberian Jay at the top of the Valtavaara fjell in winter.

Geese, Whooper Swans, Tundra Bean Geese and Taiga Bean Geese gather in surrounding fields. With good luck, Lesser White-fronted Geese, one of Finland's rarest species, can also be seen here, often feeding alongside Tundra Bean Geese, Taiga Bean Geese and Greater White-fronted Geese. Waders such as Ruff, Black-tailed Godwit and various sandpipers frequent tidal flats, while summer breeding adds further diversity.

The Liminka Bay Visitor Centre offers trails, exhibitions and observation towers with sweeping views. May and September are highlights, but Liminka Bay rewards birders year-round.

KUUSAMO VALTAVAARA

Valtavaara, east of Kuusamo near Ruka, is the highest fell in the area and a highlight of the famous Karhunkierros Trail. Rising over 490m, it offers diverse habitats including spruce forests, rocky slopes and clear ponds that support a rich boreal bird community.

The fell is prized for rare northern species such as Red-flanked Bluetail, Siberian Jay and Rustic Bunting. Forest interiors shelter Willow Tit, Three-toed Woodpecker, Hazel Grouse, Black Woodpecker, and Tengmalm's and Pygmy Owls, while the open slopes attract a variety of passerines. Spring brings a chorus of finches and warblers. Raptors can be seen, including Peregrine Falcon and occasionally Golden Eagle.

Marked trails and viewpoints provide sweeping panoramas over Kuusamo's fells and lakes, with the nearby day hut and Ruka village offering easy access. For birdwatchers and hikers alike, Valtavaara blends dramatic scenery with exceptional and diverse birdlife all year.

OULANKA NATIONAL PARK

Oulanka National Park, spanning Kuusamo and Salla near the Russian border, is one of Finland's premier birdwatching destinations. Its dramatic river valleys, pine forests, meadows and roaring rapids create a mosaic of rich northern habitats.

The taiga forests are home to classic boreal species such as Siberian Jay, Hazel Grouse, Three-toed Woodpecker and Capercaillie. Parrot and Common Crossbills feed in the pinewoods, while Golden Eagles often patrol the nearby uplands. Along the Oulankajoki River, Dippers dive in the rapids and Grey Wagtails feed on stony banks. Common Redstarts and Bramblings add colour to forest edges and meadows in summer.

During migration, the park hosts a variety of waders, waterfowl and occasional eastern vagrants. The famous Karhunkierros Trail passes through many of Oulanka's finest habitats, offering excellent opportunities for both hiking and birdwatching in one of Finland's most scenic wilderness areas.

INARI NELJÄN TUULEN TUPA

Neljän Tuulen Tupa, located in Kaamanen along Finland's E4 highway, is a renowned stop for birdwatchers exploring northern Lapland. This cozy inn and café offers a unique opportunity to observe a variety of boreal bird species from the comfort of its birdwatching-friendly facilities.

The well-stocked feeders attract species such as Pine Grosbeak, Siberian Jay, Siberian Tit, Redpoll, Brambling and Siskin. Guests can enjoy these views while sipping a warm drink in the café. The surrounding coniferous forests and wetlands also provide habitats for Willow Grouse and other northern species.

Open year-round, Neljän Tuulen Tupa offers comfortable accommodation, including rooms, log cabins and campsites, making it an ideal base for birdwatching expeditions. Its location along the E4 highway provides easy access for travellers heading north. Whether you're a seasoned birder or a casual nature enthusiast, this spot offers a memorable experience in the heart of Lapland's wilderness.

SAARISELKÄ KAUNISPÄÄ

Kaunispää, rising above Saariselkä, is one of the most easily reached fells in northern Lapland and a rewarding site for fell birdwatching. The slopes and summit offer a mix of heathland, tundra and birch woodland habitats, each supporting typical northern species.

Golden Plover display on the open tundra, their calls carrying across the fell in early summer. Meadow Pipits and Wheatears are common on the slopes, while Bluethroats add flashes of colour and song in the birch zone. The summit area is a well-known site for Dotterel, which may be seen performing their delicate courtship displays. Rough-legged Buzzards and Short-eared Owls occasionally hunt over the ridges, and Willow Grouse may appear near the trails.

Best visited from early June to mid-July, Kaunispää combines fine birding with magnificent views over the surrounding fells and valleys – a true highlight for Lapland birdwatchers.

SAARISELKÄ KIILOPÄÄ

Kiilopää, on the edge of Urho Kekkonen National Park near Saariselkä, rises to 546m and offers breathtaking views over tundra, birch forests and rolling fells. Its mix of open heathlands, rocky slopes and mountain birch groves

Willow Grouse at the Kiilopää hill.

supports a variety of characteristic Arctic bird species.

During summer, Golden Plover display across the tundra, Bluethroat sing from wet birch thickets, and Willow Grouse call softly among the heaths. Near the summit, Rock Ptarmigan blend perfectly into the stony landscape, while Dotterel may also be seen displaying on the fell plateaus. In the birch zone, warblers, Redpolls and Bramblings add sound and colour to the short Arctic summer.

The best birdwatching season is from early June to mid-July, when long daylight hours and calm weather favour exploration. With well-marked trails and magnificent scenery, Kiilopää offers a perfect introduction to Lapland's tundra birdlife.

KILPISJÄRVI SAANA

Saana Fell, rising above Kilpisjärvi in Finland's far northwest, is one of the country's most striking birdwatching destinations. The fell's slopes transition from mountain birch forest to open tundra, offering varied habitats for northern species.

On the lower slopes, Bluethroats, Bramblings and Redpolls are typical, while higher up, the tundra hosts Golden Plover, Rock Ptarmigan and occasionally Snow Buntings. Ring Ouzels nest among the rocky outcrops, adding a distinctly alpine atmosphere. Dotterel may be seen displaying on the open summit plateaus during early summer.

The brief Arctic season from June to early July brings long daylight hours and clear visibility, making birdwatching particularly rewarding. Trails from Kilpisjärvi lead through rich habitats to the summit, and nearby areas such as Malla Strict Nature Reserve add further variety. Saana combines dramatic fell scenery with special Lapland birdlife, a true highlight for birdwatchers exploring the far north.

HOW TO USE THIS BOOK

This book presents 300 bird species most likely to be encountered while birding in Finland. Most are regular breeding birds, but the guide also includes passage migrants and some vagrants that are recorded annually.

The taxonomy follows the brand-new AviList (v2025), with a few minor adjustments where species are grouped together on the same page or spread. For example, all redpolls are now treated together, so this book presents Common Redpoll and Arctic Redpoll as one species. Each species is introduced by its English and scientific names, plus the average total body length in centimetres (and wingspan for birds frequently seen in flight).

The photographs have been selected to show the plumages in which birds are most often seen in Finland. For breeding species where males and females are easily distinguished, both sexes are illustrated. Some rare vagrants may be represented by only a single image in one plumage. Winter or juvenile plumages are included where necessary. Unless otherwise noted, photographs show adult birds in which both sexes look similar.

Different morphs are also shown, such as pale and dark morphs in skuas or the brown and grey morphs of Tawny Owl. Subspecies occurring in Finland are noted as well, for example Yellow Wagtail (*thunbergi/flava*) and Nuthatch (*asiatica/europaea*).

Male – ♂
Female – ♀
Juvenile – juv.
First-winter – 1st-win.
Second calendar year – 2cy
Sub-adult – sub-ad.
Breeding – br.
Non-breeding – non-br.
Summer – sum.
Winter – win.
Autumn – aut.
Pale morph – pale
Dark morph – dark

STATUS IN FINLAND

This book describes the status of each bird species in Finland, using the following terms in the 'Where to see' section:

Resident – Species that live in Finland all year. Some may include partial migrants, where part of the population moves while others stay.

Summer visitor – Species present mainly in the breeding season and absent in winter.

Common breeder – Species that breed widely and are frequently encountered.

Uncommon breeder – Species with breeding populations in Finland but which are harder to find.

Passage migrant – Species that pass through Finland in spring or autumn, usually on their way to breeding grounds in Siberia or to wintering areas in Europe or Africa.

Rare vagrant – Species that do not normally occur in Finland but appear occasionally.

Mute Swan *Cygnus olor* 150cm, wingspan 220cm

Very large, white swan with a long, slightly curved neck. The adult has an orange bill with a black base and a prominent knob. Tail is longer than other swans. Juveniles are light brown. Mostly silent, though wingbeats in flight produce a diagnostic throbbing sound. Can be aggressive during breeding.

Where to see Resident along coastal areas, reedy lakes and the Baltic archipelagos. Often gather in large flocks outside the breeding season, with most wintering at sea if waters remain ice-free.

Bewick's Swan *Cygnus columbianus* 120cm, wingspan 180cm

Smaller than Whooper Swan, with a shorter neck, proportionally larger head and a short dark bill with less yellow. Easier to spot in flocks of Whoopers, but harder to identify when alone. In flight, wingbeats are faster and more buoyant than Whooper.

Where to see Passage migrant in Finland. Uncommonly found feeding with Whoopers in fields, mostly from southern Finland up to the Oulu region.

Whooper Swan *Cygnus cygnus* 150cm, wingspan 220cm

Large, white swan, similar in size to Mute Swan but bigger than Bewick's Swan. The neck is normally straight and slender. Legs are black, and the wedge-shaped bill has a large yellow area extending to the base. The head can appear brownish from feeding in water. Juveniles are grey, but bill shape and size are like adults. Very vocal, producing loud bugling calls. Can be aggressive during breeding.

Where to see National bird of Finland. Common breeder. Breeds widely across the country in swamps, wetlands and small lakes. Migratory, but some winter along the southern Baltic coast. Large flocks gather in fields during migration.

Brant Goose *Branta bernicla* 58cm, wingspan 110cm

Small, dark goose, slightly smaller and shorter-necked than Barnacle Goose. Head, neck and breast are black, with a pale collar at the neck. The back and belly are darker than Barnacle, making it easy to spot in flight during migration. Noisy in Arctic migration, giving a high-pitched, rapidly repeated nasal *rrrrot-rrrrot* call.

Where to see Spring passage migrant along Finland's southern coast in large flocks, occasionally mixed with other geese. Small numbers occur in autumn.

Red-breasted Goose *Branta ruficollis* 57cm, wingspan 115cm

Slightly smaller than Brant Goose and Barnacle Goose. Distinctive at close range, with a red breast and cheeks contrasting with a black and white body. In flight, it shows a short neck, black underwings, dark upperwings with two white bars and a black belly. Can be difficult to spot from a distance within Barnacle or Brant flocks.

Where to see Uncommon passage migrant along Finland's southern coast in spring, usually among Barnacle or Brant flocks. Rare during autumn migration.

Canada Goose *Branta canadensis* 90cm, wingspan 170cm

Larger than all other goose species. A long neck, thinner white cheek patch and brown back help distinguish it from Barnacle Goose. With a black neck and white cheek, it is easy to spot among flocks of Greylag, Taiga and Tundra Bean Geese. Call is a deep, resonant honk, *ah-honk*.

Where to see Common breeder in southern Finland, mostly along the coast and at large inland lakes. Less conspicuous on migration than other geese but can join mixed flocks.

Barnacle Goose *Branta leucopsis* 65cm, wingspan 130cm

Smaller and shorter-necked than Canada Goose, and slightly larger than Brant Goose and Red-breasted Goose. A large white cheek patch continues to the forehead, and the grey back helps distinguish it from Canada. In flight, white cheek and pale belly make it more distinctive than Brant. Call is a rolling *ka-ka-ka*.

Where to see Locally common breeder in southern Finland, mainly around Helsinki. Common spring migrant in large flocks along the south coast, also migrating in autumn and feeding in eastern fields.

Bar-headed Goose *Anser indicus* 75cm, wingspan 150cm

Medium-sized goose, easily identified by pale grey body, black wing feathers and neck, and white head with two black transverse stripes at the rear.

Where to see European population originates from captive birds, with most sightings being their descendants. Uncommon yearly migrant, mostly along the coast but can appear anywhere from spring to autumn. Often moves in flocks with other goose species.

Greylag Goose *Anser anser* 80cm, wingspan 158cm

Large, robust goose with brown-grey plumage and a bright orange bill. Pale head and pink legs distinguish it from Taiga Bean Goose. Undertail is white. In flight, the upperwings are broadly pale blue-grey, a very noticeable feature. The call is a loud, nasal *aangh*.

Where to see Fairly common breeder along the coast, mostly in Åland and southern archipelagos. Migrates in small or mixed flocks with other geese. During spring and autumn, also feeds in fields but prefers swimming in water.

Lesser White-fronted Goose *Anser erythropus*

60cm, wingspan 125cm

Like Greater White-fronted Goose but smaller, with a shorter pinkish bill, prominent yellow eye-ring, and white blaze reaching the forecrown. Usually shows less prominent belly marking. In flight, neck is shorter, head smaller, wings narrower and wingbeats faster than Greater White-fronted.

Where to see Rare spring passage migrant, mostly around the Oulu area. Often feeds with other geese in fields. A few breeding pairs may occur in northern Lapland swamps.

Greater White-fronted Goose *Anser albifrons*

70cm, wingspan 145cm

Smaller than Greylag Goose and Taiga Bean Goose. White forehead shield and pinkish bill are distinctive. Black patches on the belly and orange legs are visible in flight. In large mixed flocks, belly stripes make it easier to identify. Call is a high-pitched, laughing *kla-ha-ha*.

Where to see Passage migrant, mostly in spring but also in autumn. Gathers in large flocks on fields, often alongside Tundra and Taiga Bean Geese and other geese species.

Tundra Bean Goose *Anser serrirostris* 75cm, wingspan 158cm

Similar to Taiga Bean Goose but slightly smaller, with a shorter neck. The head is darker, and the bill is shorter, more triangular and less orange than in Taiga Bean. Occasionally resembles Pink-footed Goose but bill and feet are orange, and back is darker.

Where to see Passage migrant. Large flocks stop to feed in fields during migration to Siberia.

Pink-footed Goose *Anser brachyrhynchus* 70cm, wingspan 150cm

Smaller than Taiga Bean Goose, with a short pinkish bill and pink legs. Dark brown head, short neck and blue-grey back help distinguish it from Taiga Bean. In flight, it shows a broad white tail tip, pale upperwing and noticeably short neck.

Where to see Uncommon passage migrant, mostly feeding in fields with Tundra and Taiga Bean Goose flocks from southern Finland up to the Oulu region. Rare in Lapland.

Taiga Bean Goose *Anser fabalis* 78cm, wingspan 158cm

Medium-sized goose, smaller than Greylag Goose and Canada Goose but larger than most other geese. The head is darker than Greylag, with a smaller bill featuring an orange tip and black base. The back is dark with pale edges on the wing coverts, and the legs are orange. In flight, it shows dark upperwings and a narrow white tail-tip band. Call is a deep, nasal honk, *aangh*.

Where to see Uncommon breeder in the north, at swamps and small lakes. Common passage migrant in the south, often feeding in large flocks on fields and usually mixed with other geese species.

Mandarin Duck *Aix galericulata* 45cm, wingspan 70cm

Male is unmistakable in breeding plumage, with shiny multicoloured feathers, orange head plumes, a broad whitish band from the bill over the eye to the crest, orange sails on the back, and a small pink bill. The female is plain olive-grey with a white eye-stripe and line behind the eye. After breeding, males moult into a female-like eclipse plumage, as in other ducks.

Where to see Uncommon annual visitor, mostly in the south but possible anywhere. Prefers well-vegetated coastal bays, small lakes and ponds.

Ruddy Shelduck *Tadorna ferruginea* 65cm, wingspan 120cm

Slightly larger than Common Shelduck, with a long neck and legs. The body is orange-brown, contrasting with a paler head, especially around the eye. Bill and legs are black. Males in breeding plumage show a thin black neck-collar. In flight, striking white wing panels contrast with dark flight feathers, making it distinctive.

Where to see Rare vagrant in Finland, with only a few records each year. Usually seen alone on large lakes or coastal bays in the south.

Common Shelduck *Tadorna tadorna* 60cm, wingspan 110cm

A large, goose-like duck, unmistakable with its bold, mostly white plumage with dark green head and broad, rusty-brown breast-band encircling the body. The bill and legs are pink, and males show a prominent red bill knob during the breeding season. Juveniles are duller, white below and grey-brown above. In flight, displays striking white wings with bold black markings.

Where to see An uncommon but regular breeder from the south coast to the Oulu region, mainly along coastal shores and lagoons. After breeding, large moulting and post-breeding flocks may form in sheltered bays.

♂ ♀

Long-tailed Duck *Clangula hyemalis*

43cm (+10–15cm central tail feathers in male), wingspan 74cm

A small sea duck with a distinctive long tail in males. Plumage is mottled black, white and brown, with a rounded head and short bill. Plumage colouration varies through the year with seasonal changes. In flight, white lines on the back or sides of the rump are visible. Very vocal; call is a loud, high-pitched nasal yodeling *ow-ow-delee*.

♂ sum.

Where to see Uncommon breeder. Breeds in small numbers along the Baltic Sea coast and on lakes in northern Lapland. Winters in Baltic Sea. Migrates in large flocks, especially during spring migration to the Arctic.

♂ ♀ win.

Steller's Eider *Polysticta stelleri* 45cm, wingspan 72cm

♂

♀

A small sea duck with striking, colourful plumage. Males in breeding plumage display a unique mix of black, white, iridescent green, and chestnut-orange breast and flanks, with a pale blue-grey back, whitish head and black spot on the neck. Females are mottled dark brown and are more subdued. In flight, females show bluish wing coverts with white outer edges.

Where to see Uncommon visitor during spring and autumn migration on the coast and in the archipelago.

King Eider *Somateria spectabilis* 59cm, wingspan 92cm

Male is unmistakable. The head is strikingly multicoloured, with an orange knob at bill base, pinkish bill, blue-grey face and greenish cheeks. Black-and-white body. Females are mottled brown with a pale face, resembling females of Common Eider but appearing shorter-billed, rounder-headed and more of a warm brown colour.

♂

Where to see Uncommon visitor during spring and autumn migration, usually along the coast and in the outer archipelago. Most often seen in mixed flocks with Common Eiders.

♀

Common Eider *Somateria mollissima* 65cm, wingspan 100cm

A large sea duck with a heavy body and thick, sloping bill. Males in breeding plumage are black and white with a greenish nape, while females are mottled brown and well-camouflaged. In flight, it appears chunky with a long neck. The male's call is a soft, cooing *ah-ooooh*.

♀♂

Where to see A common breeder on rocky islets and small islands along the Baltic coast. Males form large flocks at sea once females begin incubation.

♀♂

Common Scoter *Melanitta nigra* 49cm, wingspan 77cm

Slightly smaller than Velvet Scoter. The male is almost entirely black with a distinctive bulbous knob at the base of the bill and a yellow-orange patch near the nostrils. The female is dark brown with a large pale cheek patch extending well down the neck and a slimmer bill. In flight, it appears uniformly dark.

Where to see Uncommon breeder. Breeds mainly on lakes in Lapland. Often seen around Baltic Sea islands and during spring migration to the Arctic, when large flocks pass along the coast.

♂ ♀

♂ ♀

Velvet Scoter *Melanitta fusca* 54cm, wingspan 87cm

A large, heavy sea duck with a bulky body and rounded head. Males are mostly black with a bold white patch beneath the eye, a broad white wing speculum and an orange bill with a black knob. Females are dark brown with pale cheek patches and a duller bill. In flight, both show a distinctive white panel on the wings.

Where to see A fairly common breeder along the Baltic Sea and at large northern lakes, especially north of Oulu.

♂

♀

♀♂

Goldeneye *Bucephala clangula* 44cm, wingspan 70cm

A medium-sized diving duck with a compact body and rounded head. The male has a glossy greenish-black head with a bright white oval spot near the bill, striking yellow eyes and a mostly white body with black back and wings. The female is grey with a chocolate-brown head, white collar and yellow eyes. In flight, the short wings flash bold white patches, and rapid wingbeats create a distinctive whistling sound.

Where to see A widespread and abundant breeder throughout Finland, from the southern archipelago to northern Lapland, inhabiting lakes, rivers and sheltered coastal bays.

♂

♀

Smew *Mergellus albellus* 41cm, wingspan 62cm

A small, handsome duck a little smaller than a Goldeneye. The male is mostly white with bold black markings in front of the eye and on the nape, back and sides; greyish flanks. The female has a grey body and a reddish-brown head with a white cheek and throat. In flight, the wings are mostly black with prominent white patches.

Where to see Summer visitor. A shy species, breeding mostly in Lapland. Prefers shallow lakes with abundant vegetation but also occurs on more barren lakes, river backwaters and bog pools. Migrates through Finland and can be seen across the country during migration.

Red-breasted Merganser *Mergus serrator* 55cm, wingspan 75cm

Smaller and slimmer than Goosander but similarly shaped. The male is striking, with a glossy green head adorned with a shaggy crest, a white collar, a brown-streaked breast, grey sides and a black back. It has red eyes and a slender, slightly upturned red bill. The female is duller, with a light brown head and a less-defined border to the neck. Both sexes show a dark neck and white wing patches with black lines in flight.

♂

♂ ♀

Where to see An uncommon breeder found across Finland, especially along coasts, large clear lakes and rivers. Moves southward in winter as inland waters freeze.

Goosander *Mergus merganser* 63cm, wingspan 86cm

A large, elegant diving duck with a long, slender body and a red, hooked bill. The male is striking with a dark green head, white body and black back. The female is pale grey with a reddish-brown head, short crest and sharply defined white chin. In flight, both sexes show bright white wing patches, and the long neck gives a graceful profile. Dives deeply to catch fish.

♂

♀

Where to see Common wide-spread breeder across Finland, breeding along coasts, islands, rivers and clear lakes. It gathers in large flocks and moves south in winter as waters freeze.

Red-crested Pochard *Netta rufina* 55cm, wingspan 88cm

♀ ♂

A medium-sized diving duck with a rounded head and stout body. The male has a striking chestnut-red head, contrasting with a pale grey back and black breast and underparts. The eyes and bill are bright red. In flight, the white wing pattern is distinctive. Females are mottled brown with a paler face and lack the male's vivid colours.

Where to see Uncommon visitor, mostly in the south but occasionally found farther north. Prefers sheltered coastal bays and small, lush inland lakes; often seen among other duck species.

Common Pochard *Aythya ferina* 45cm, wingspan 71cm

A medium-sized diving duck similar in size to Tufted Duck. The male has a rich chestnut-red head, black breast, pale grey body and bright red eyes, with a bluish-grey bill tipped in black. The female is brownish and duller overall, with dark eyes and a less contrasting pattern. In flight, this species lacks the bold white wing-bars of Tufted and Scaup.

♂

Where to see An uncommon breeder in southern Finland, rarer further north. Prefers lush lakes with both open water and dense vegetation, avoiding small ponds.

♀

Tufted Duck *Aythya fuligula* 45cm, wingspan 69cm

A medium-sized diving duck. The male has a distinctive tuft of feathers on the head, glossy black plumage with a bluish sheen, white flanks and bright yellow eyes. The female is dark brown with a smaller tuft and duller eyes. Both have a blue-grey bill with a broad black tip. In flight, white wing-bars stand out. Calls are harsh *karr-karr*.

Where to see Fairly common throughout Finland. Found in archipelagos, coastal bays, lush inland lakes and freshwater bodies. Often in flocks; some overwinter in the south if waters remain unfrozen.

♂

♀

Greater Scaup *Aythya marila* 47cm, wingspan 76cm

Similar to Tufted Duck but with a rounded head and no tuft. The male has a greenish sheen to the head and a whitish back. The female is greyish-brown with a large white patch at the bill base, though this varies. Both have yellow eyes and a blue-grey bill with a small black tip. In flight, wing-bars resemble those of Tufted Duck.

Where to see Uncommon breeder in the Baltic Sea archipelago and parts of northern Lapland. More frequent during migration, often on freshwater lakes and coasts, usually with Tufted Ducks.

♂

♀

Garganey *Spatula querquedula* 40cm, wingspan 63cm

A small, Teal-sized duck. The male in breeding plumage is striking, with a dark brown head with a strong white crescent above the eye, a brown breast and a grey body. The female is brown with strong head stripes and a whitish throat, a large grey bill and subtle patterns. In flight, males show pale grey forewings and two bold white wing-bars. After breeding, males moult into female-like plumage.

Where to see Uncommon breeder in southern and central Finland. Shy and well-camouflaged, often hard to spot. Prefers lush lakes, wetlands and vegetated coastal bays.

Shoveler *Spatula clypeata* 50cm, wingspan 78cm

The long, broad bill, usually tilted downwards, is its most distinctive feature in all plumages. Slightly smaller than a mallard with a shorter neck, it can appear compact. Breeding males are striking, with a green head, yellow eye, white breast and chestnut belly and flanks. Females and juveniles resemble brown Mallards but have the signature large bill and short neck. Males moult into female-like plumage after breeding.

Where to see Relatively scarce breeder, mainly along the coastline and inland in southern and central Finland. It is more common during migration in the south, often gathering in flocks with other ducks in lush lakes, coastal bays, wetlands and marshlands.

♂

Gadwall *Mareca strepera* 50cm, wingspan 84cm

Slightly smaller and more elegant than Mallard, the male in breeding plumage has a light brown head, mostly grey body and black stern. Females resemble Mallards but have an orange-sided bill with a dark central line, a whitish throat and occasional white wing patches. Unlike Mallard, the blue wing patch is absent. In flight, males show black and white patches and chestnut in the upper wing; their whitish belly is a key difference. Males moult into female-like plumage after breeding.

Where to see Uncommon summer visitor, mainly in southern Finland and along the coast to the Oulu region. Breeds in lush lakes, ponds, wetlands and marshlands.

♂

♀

♀♂

Wigeon *Mareca penelope* 45cm, wingspan 78cm

A medium-sized duck with a large, rounded head and a small, light grey bill with a black tip. The male in breeding plumage has a chestnut head and neck, a creamy crown-stripe, pink breast, grey body and a black-and-white stern. The female is brown with rufous flanks. In flight, males show large white patches on the upper wing and a white belly in all plumages. Males moult into female-like plumage after breeding. The call is a distinctive loud whistle.

Where to see Fairly common breeder throughout Finland. Prefers bays, lakes, ponds with vegetation and coastal meadows. Often forms large flocks with other ducks such as Teal and Mallard.

Mallard *Anas platyrhynchos* 55cm, wingspan 88cm

Bigger than Teal, Mallard is a common and widespread dabbling duck. The male in breeding plumage is easily recognised by its glossy green head, white neck-ring and chestnut-brown chest. Females are brown with an orange bill marked by dark patches. Both sexes have a distinctive blue speculum bordered with white, visible in flight. Males moult into female-like plumage after the breeding season, like all ducks.

Where to see Common everywhere in Finland. Found in lakes, rivers, wetlands, coastal waters and urban parks. Some overwinter in Finland when water bodies remain unfrozen.

Pintail *Anas acuta* 55cm (+10cm tail in male), wingspan 83cm

This dabbling duck has a longish neck and remarkably long tail. The male in breeding plumage is easily recognised by its dark brown head, light blue bill with a black line on top, white breast and lower neck and narrow white stripes extending into the brown head. The female is paler brown than most ducks, with a grey bill and long pointed tail. In flight, the male shows a white trailing edge of the wing with a blackish-green speculum bordered rufous at the front. The long, narrow neck is also visible. Males moult into female-like plumage after breeding.

Where to see Uncommon breeder throughout Finland. Prefers bogs, fens, fen banks and flooded coastal meadows. Gathers with other ducks during migration.

Teal *Anas crecca* 35cm, wingspan 56cm

♂

A small dabbling duck with a compact body and short neck. The male in breeding plumage has a chestnut head with a bold, iridescent green eye patch extending from the eye to the back of the head. The body is grey with a creamy-white undertail. Females are mottled brown with a subtler facial pattern. Both sexes show a bright green speculum bordered by white in flight. Males moult into female-like plumage after breeding. The male's call is a clear, whistled *crii crii*.

Where to see A common breeder throughout Finland. Found in bays, lakes, wetlands, rivers and coastal archipelagos, often in mixed flocks.

♀ ♂

Rock Ptarmigan *Lagopus muta* 33cm

♂ sum.

♀ sum.

Smaller and greyer than Willow Ptarmigan. In summer plumage, it shows a greyish head, neck, and back with white wings and belly, and a black tail. The male has a red eyebrow comb, typical of all male grouse. In winter, both sexes turn almost completely white, except for a black tail, and the male retains black lores. The call is a dry, snoring *arr ka-karr.*

Where to see A resident species of the mountains of Lapland, breeding high in barren, rocky terrain with sparse vegetation. Perfectly camouflaged, it is often detected only when it moves or calls.

♀♂ win.

Willow Ptarmigan *Lagopus lagopus* 39cm

A medium-sized grouse with a distinctive seasonal plumage. In summer, the male has a reddish-brown head and neck contrasting with a white body, while the female is more uniformly brown with white underparts. Both sexes show a red eyebrow comb in all plumages. In winter, they turn almost completely white except for the black tail. In flight, they appear mostly white. During the breeding season, the male is very vocal, producing loud, comical, laughing calls.

Where to see A resident species breeding mainly in northern Finland's swamps and bogs. In Lapland, found on the lower mountain slopes with denser vegetation, unlike Rock Ptarmigan.

♂ br.

aut.

win.

Capercaillie *Tetrao urogallus* 82cm (incl. 25cm tail)

A large and majestic grouse, the male Capercaillie is instantly recognisable by its fan-shaped tail, dark greenish-black plumage, brown wings, white shoulder patch and bright red eyebrow above the eye. The pale bill and upright tail are distinctive during courtship displays. The female is larger than a hen Black Grouse, with a thick neck, large head and more colourful plumage, notably an orange breast, greyish head and back, white belly and finely barred patterning.

Where to see A resident of mature coniferous forests, fairly common in Kuusamo and eastern Finland. Often seen feeding along forest roads or perched in trees.

♂

♀

♂ ♀

Black Grouse *Lyrurus tetrix* 54cm

Smaller than Capercaillie, the male Black Grouse is unmistakable with its glossy blue-black plumage, lyre-shaped tail and white undertail feathers shown prominently during display. Bright red eye combs and bold white wing-bars, visible both at rest and in flight, add to its striking appearance. Females are smaller, mottled brown with a short neck and small head, well-camouflaged in vegetation. Courtship is spectacular, with males gathering at leks in open areas, producing bubbling, gurgling and hissing calls.

Where to see A resident throughout Finland, favouring forest edges, clearings, fields and swamp borders. Often seen feeding in birches and willows.

♂

♀

♀♂

Hazel Grouse *Tetrastes bonasia* 37cm

♂

♀

A small, plump grouse with a short neck and small head. Upperparts are greyish with brown wings, and underparts are whitish with dark brown spotting on the belly. The male has a distinctive black throat patch. In flight, the wings beat rapidly, and a narrow black band is visible on the tail. The call is a high, thin *tsii-iiih ti-ti-ti-ti*.

Where to see A resident in mature spruce forests. Often secretive, walking quietly on the ground or perching low in trees. Occasionally seen along roadsides feeding on buds.

Grey Partridge *Perdix perdix* 30cm

A small game bird, about the size of a Wood Pigeon. It has a rounded, ball-like appearance due to its large head, short neck and short tail. The head and throat are orange-yellow, the body mostly grey with reddish-brown streaks on the sides and a brown patch on the belly.

Where to see Resident from southern Finland up to the Oulu region. Most often found in farmland and fields near villages, usually in small flocks. Easiest to spot in winter when contrasting against snow.

♀♂

Pheasant *Phasianus colchicus* 80cm (incl. 40cm tail)

♂

♀

Male unmistakable with its colourful plumage and exceptionally long tail. The body is orange-brown with black belly stripes, a large red wattle over the blackish-green head, a white neck ring and a pale bill. The female is light brown with dark spots on the back. Often walks or runs on the ground. The male's call is a loud, rasping, croaking *koor-kok*.

Where to see Resident and fairly common from southern Finland up to the Oulu region. Prefers human-modified landscapes and is often found in urban areas, parks and gardens.

Quail *Coturnix coturnix* 17cm

A tiny bird with a short neck and tail, resembling a young Pheasant. Plumage is buff-brown with black markings on the back and a dark throat, providing excellent camouflage in grassy fields. The male gives a distinctive, repetitive, high-pitched *but but-ut* call, most often heard at dawn or dusk.

Where to see A summer visitor and rare breeder in southern farmlands. Very secretive and difficult to observe, usually moving quietly on the ground. Occasionally glimpsed along field edges or near roads.

♂

Little Grebe *Tachybaptus ruficollis* 26cm

sum.

win.

This tiny grebe is an active diver. Its small size and short neck give it a compact, dumpy appearance. In breeding plumage, it has a black head, rich chestnut cheeks, dark back and a dark bill with a yellow base. In winter, it turns duller brown with a pale throat and buff body. Like all grebes, it often disappears underwater for long periods.

Where to see Uncommon visitor. Occasionally breeds in the south on small, well-vegetated ponds and lakes; otherwise found on lakes or coastal waters.

Slavonian Grebe *Podiceps auritus* 35cm

A medium-sized grebe and full-time diver. In breeding plumage, it shows a black head with striking orange-yellow ear-tufts behind the bright red eyes, a black back and chestnut flanks and neck. The bill is short and black. In winter, the plumage is plainer, with a white face and dark grey crown.

Where to see Summer visitor breeding on wetlands, ponds and well-vegetated lakes, also in shallow coastal bays. Found from the south to the Oulu region; rare further north.

sum.

Red-necked Grebe *Podiceps grisegena* 43cm

Slightly smaller than Great Crested Grebe, with a shorter rusty-brown neck. It has a black head with pale grey to whitish cheeks, a black back and a black bill with a yellow base. Like all grebes, it dives frequently and stays underwater for long periods. In the breeding season, it is noisy and easily recognised by its harsh, nasal, rhythmic *kree-kree-kree* call.

sum.

Where to see Locally common breeder from the south to southern Lapland. Prefers larger, well-vegetated lakes and sheltered coastal bays.

Great Crested Grebe *Podiceps cristatus* 48cm

The largest grebe, with a long slender neck, dark red eyes and a pinkish bill. In breeding plumage, it is unmistakable with its striking black crest, orange-brown cheek tufts and white face. The body is mostly dark above and pale below. In winter, it appears plainer, with a grey head and back, and white cheeks and breast. Pairs perform an elaborate and elegant courtship display in spring.

Where to see Common breeder on large lakes, ponds, and sheltered coastal lagoons, preferring calm, shallow waters with reedbeds or floating vegetation. Most common from the south to the Oulu region, scarcer further north.

sum.

win.

Common Cuckoo *Cuculus canorus* 34cm

juv.

A hawk-like bird with long, pointed wings and tail. The male has grey upperparts, white underparts with fine black barring and black-and-white striped underwings. Females are either grey or reddish-brown. Both sexes have yellow eyes with a black pupil and a short, slightly down-curved beak. Its flight is direct and steady, unlike raptors. Most often recognised by the male's unmistakable *cuc-koo* call.

Where to see A summer visitor, commonly heard in late spring and summer in conifer-dominated forests, clearings and farmland edges, or during migration in the archipelago.

Turtle Dove *Streptopelia turtur* 27cm

The smallest dove, more variegated in colour than other species. It has a grey head, pinkish breast, orange-spotted back, bright red eyes with red orbital skin, a black bill and a distinctive black-and-white striped patch on the neck. In flight, shows ash-grey underwings, black upper flight feathers, a white belly and a short tail with a striking white-edged pattern. Its deep purring call, *turrrrr-turrrrr-turrrrr*, is rarely heard in Finland.

Where to see Uncommon visitor in southern Finland, appearing during migration in the archipelago, in fields, meadows and occasionally urban centres.

Collared Dove *Streptopelia decaocto* 31cm

A medium-sized dove, pale brown overall, with a narrow black collar edged in white across the nape, red eyes and a black bill. In flight, shows black flight feathers and a long tail with broad white tips on the upperside. The call is a soft, rolling, three-syllable *coo-coo-coo*.

Where to see Uncommon breeder in the south and along the west coast up to the Oulu region. Prefers human settlements, typically nesting near rural homes, urban yards and in urban centres.

Wood Pigeon *Columba palumbus* 41cm

A large, grey pigeon with a white neck-patch and a pinkish breast. Pale eyes with dark pupils help distinguish it from Stock Dove. In flight, it shows bold white wing-bars, black flight feathers and a black tail with a white terminal band. Its flight is heavy, giving a pot-bellied appearance. The call is a deep, owl-like hooting, typically five syllables: *hooh-hoo-hoo-hoo-oo*.

Where to see Common breeder throughout Finland except in northeastern Lapland. Found in varied habitats including urban yards, parks, city centres, farmland and forests.

Rock Dove / Feral Pigeon *Columba livia* 32cm

Similar to Stock Dove. Overall grey with a darker head, reddish eyes, black bill and bold black wing-bars. In flight, shows a dark head contrasting with the light grey body, a white rump patch and a narrow black tail-band. Feral Pigeons, descended from domestic Rock Doves, show a wide variety of plumage colours and patterns.

Where to see Common in towns and cities, foraging in parks, streets and squares. Nests on buildings, bridges and other urban structures.

Feral Pigeon

Stock Dove *Columba oenas* 30cm

Smaller and more compact than Wood Pigeon. Overall grey, with the head and back of similar tone. The breast is pinkish, and the neck has a shiny green patch. The bill is reddish with a white tip, and the eyes are black. In flight, shows a wide black wing-stripe, a broad black tail-band, and two small black wing patches. The underwing appears darker than in Wood Pigeon. Call a soft, monotonous *ooo-ue*.

Where to see Uncommon breeder in the south up to the Oulu region, mostly in coastal areas. Breeds in agricultural landscapes, nesting in tree hollows at forest edges and feeding in nearby fields.

Crane *Grus grus* 107cm, wingspan 200cm

A huge, elegant bird with very long legs and neck. Plumage pale bluish-grey with black and white markings on the head and upper neck, and a red crown. Bill long and pointed; tertials elongated and shaggy. During breeding, the back may appear brown-stained from swamp soil. Sexes alike; juveniles have a brownish head. In flight, the neck is held straight, unlike Grey Heron, and the flight feathers are darker than the coverts. Calls are a loud, trumpeting *kkruu-ih kraw*, while flying birds utter a rolling *krro*.

Where to see Summer visitor and common breeder in swamps, wetlands and lake shores. Often feeds in fields and migrates in large, V-shaped flocks.

Water Rail *Rallus aquaticus* 25cm

juv.

Smaller than Moorhen, with a rounded head, long neck and compact body. Bill long, slightly down-curved and red. Upperparts brown with black streaks; face and breast bluish-grey; flanks boldly barred black and white. Tail often cocked, showing a creamy undertail. Legs pinkish, eyes red. Juvenile duller, without grey tones or red bill. Usually heard rather than seen, its call is a piglet-like squeal *gruiit-grroit-grui-gru*.

Where to see Uncommon and very secretive breeder in southern Finland, inhabiting reedbeds of lakes, bays and wetlands. Occasionally appears at reed edges.

Corn Crake *Crex crex* 24cm

A partridge-like bird but slimmer and shorter. Compact, rounded body, small head, short sturdy bill and relatively long neck give it a distinctive look. Face, throat and chin greyish; belly and back heavily striped. Wings warm reddish-brown, and in flight the long dangling legs are conspicuous. Call unmistakable: a loud, rasping *crex-crex-crex*.

Where to see Summer visitor in southern Finland. Prefers fields, meadows and marshes near wetlands. Very secretive, most often heard rather than seen; sometimes glimpsed at field edges or roadsides.

Spotted Crake *Porzana porzana* 21cm

Smaller than Water Rail, with a compact body, short neck and short bill. Upperparts dark brown, heavily streaked; face and breast greyish with a dark patch between eye and bill. Bill yellow with a reddish base. Breast and belly white-spotted; undertail coverts buff. Legs olive-green. Call a high-pitched, rhythmic *huitt-huitt-huitt-huitt*.

Where to see Summer visitor, secretive and hard to observe. Breeds in reedbeds and sedge meadows from southern Finland up to the Oulu region, rarer further north; occasionally glimpsed at reed edges.

Little Crake *Zapornia parva* 18cm

♀

♂

Smaller than Spotted Crake, with a Water Rail-like shape. Face, breast and belly bluish-grey. Bill is short and thin, olive-green with a reddish base; eyes bright red. Tail often cocked, revealing black-and-white barred undertail coverts. Males have bluish-grey underparts and dull brown back; females are buff-brown below with brown back. The song, usually at night, starts with slow *kua*… *kua*… notes, accelerating into a rapid series.

Where to see Uncommon visitor in southern Finland. Secretive; found in reedbeds, sedge meadows and sheltered wetlands. Does not breed in Finland but a few appear annually.

Moorhen *Gallinula chloropus* 29cm

juv.

Smaller than Coot but similarly plump. Overall greyish with a dark brown back and distinctive white flank lines. Bill and frontal shield bright red with a yellow tip. Tail often cocked, showing white undertail coverts with a black central stripe. Juveniles are grey-brown with a dirty-white throat and similar flank and undertail markings. Legs long and olive-green. Voice loud and harsh, a repeated *kurrk*, especially when alarmed.

Where to see Uncommon breeder in southern Finland. Found in wetlands, ponds and reedbeds, occasionally venturing into open water or grassy edges.

Coot *Fulica atra* 39cm

Plump, broad-bodied waterbird, mostly blackish with dark red eyes. The head is small and rounded, with a white bill and prominent frontal plate. Legs are greenish-grey with long, lobed toes, ideal for swimming and walking on floating vegetation. Juveniles are duller, with greyish plumage and a paler bill. In flight, wings are broad and rounded. Call is harsh and grating, often a gargling *kyorrrl*.

Where to see Summer visitor. Common on lakes, ponds, marshes and slow rivers throughout Finland. Often forms large flocks in open water.

Avocet *Recurvirostra avosetta* 44cm

Unmistakable, elegant and slender wader with striking black-and-white plumage. It has a long, thin, upcurved black bill and long, pale blue-grey legs. Crown, nape and neck are black, contrasting with white body. In flight, it appears mostly white with bold black wing-bars and wing-tips.

Where to see Uncommon visitor in the south. Prefers shallow coastal beaches, bays and wetlands. Most often seen wading in lagoons, often alongside other wader species.

Oystercatcher *Haematopus ostralegus* 42cm

Large, compact wader with striking black-and-white plumage. It has a long, thick, bright orange bill, red eyes and pinkish legs. Head, wings and tail tip are black; belly, lower back, uppertail and bold wing-stripe are white. In flight, the black-and-white pattern is very distinctive. Voice loud and clear, a high-pitched *ku-biik*.

Where to see Summer visitor. Breeds along the Baltic Sea coastline and on islands in the archipelago. Rare inland.

Grey Plover *Pluvialis squatarola* 28cm

Slightly larger, chunkier and more robust than Golden Plover, with a stronger bill. Overall, it appears greyer in all plumages. In breeding plumage, the face, chest and belly are black, bordered by a white stripe. The back is mottled grey, and both bill and legs are black. Juvenile and winter plumages are much greyer overall. In flight, it shows a prominent white wing-stripe, black underwing armpits and a white upper tail.

Where to see Passage migrant. Most often seen feeding on beaches, along the coastline and on islands of the Baltic archipelago during spring and autumn migration. In spring, large flocks migrate through Finland to Siberia.

br.

Golden Plover *Pluvialis apricaria* 27cm

juv.

juv.

Medium-sized wader with a short, thin black bill. In breeding plumage, the back is mottled black-brown, while the face, breast and belly are mostly black with a distinct white border. Juveniles are light brown with a pale spot near the ear. In flight, white underwings (armpits) and a narrow white wing-stripe are visible.

Very vocal during the breeding season; call is a loud, plaintive whistle *piiiy*.

Where to see Summer visitor. Breeds in open swamps throughout Finland, including the highlands of Lapland. During migration, gathers in large flocks on fields and coastal meadows.

br.

Dotterel *Eudromias morinellus* 22cm

Smaller than Golden Plover. In all plumages, a strong white eyebrow and pale breast-band are visible, and the bill is short and very thin. In breeding plumage, the back and breast are greyish-brown, with a distinctive reddish-brown belly as the key feature. Juvenile and winter plumages are mostly creamy-buff with a pale breast-band and broad white supercilium. In flight, the upperwing is mostly uniform with a narrow white trailing edge. The call is a soft, rolling *pyurr*.

Where to see Uncommon breeder in the highlands of Lapland. During spring and autumn migration occurs in fields across Finland.

♂ br.

♀ br.

Ringed Plover *Charadrius hiaticula* 18cm

A compact plover. Features a black-and-white facial mask, broad black breast-band extending to the nape, and a white neck with a distinct white stripe above the eye. Bill short and orange with a black tip; legs also orange. The juvenile has a stout bill, broad white supercilium and brown facial mask. In flight, it shows prominent white wing-bars. Call is a clear, plaintive, two-note whistle, *too-eet*.

Where to see Breeds mainly along southern coasts and archipelagos, occasionally inland north to the Kuusamo region.

Little Ringed Plover *Thinornis dubius* 17cm

Similar to Ringed Plover but smaller and more delicate. Key features include a bright yellow orbital ring, thin black bill, pinkish legs and more white on the forehead above the black mask. Slimmer body with longer legs. Juvenile has a slender bill, faint buff supercilium, narrow orbital ring and pale brown facial mask. In flight, lacks the prominent white wing-bars of Ringed. Call is a soft, whistled *peew*.

Where to see Summer visitor. Breeding sparsely up to central Lapland. Prefers gravel pits, storage yards, industrial areas and muddy natural shores.

Lapwing *Vanellus vanellus* 30cm

Medium-sized wader with a distinctive black face, head and breast, and a long, thin crest. The back shines dark green with a purplish gloss, contrasting with the white belly. Bill is short and black, and legs are pinkish. Wings are broad and rounded, and flight is deliberate and flappy. Performs tumbling flight displays in breeding season. The call is a characteristic, cat-like mewing *pee-wit*.

Where to see Common breeder up to the Oulu region, rarer in Lapland. Prefers open bogs, wet meadows, grass fields and pastures, often nesting in loose colonies.

Kentish Plover *Anarhynchus alexandrinus* 16cm

Tiny, delicate plover, paler than Ringed Plover. Bill black, legs black to greyish. In breeding plumage, the male shows an orange neck with black markings on the sides, a black forehead stripe and eye-stripes extending both in front and behind the eyes. Females, juveniles and non-breeding adults are pale grey-brown with a white neck-line and subtle brown side stripes.

Where to see Rare vagrant, annual visitor along the Baltic Sea coast, occasionally found inland on muddy banks and wader-rich habitats.

Whimbrel *Numenius phaeopus* 41cm

Smaller than Curlew, with a noticeably shorter bill. It has a distinctive head pattern with a bold crown stripe and clear eye stripes. The plumage is otherwise similar, with a wedge-shaped white rump visible in flight; wingbeats are faster than Curlew's. The call is a loud, rapid whistle, *pu-pu-pu-pu-pu*, often heard in flight.

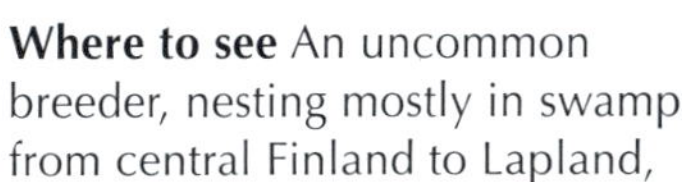

Where to see An uncommon breeder, nesting mostly in swamps from central Finland to Lapland, and occasionally in open fields. During spring migration, it can appear almost anywhere in Finland.

Curlew *Numenius arquata* 53cm

Largest wader, unmistakable with its very long, down-curved bill. It can only be confused with Whimbrel, but Curlew is larger and has a longer bill. Plumage is streaked grey-brown with barred underparts. In flight, it shows a wedge-shaped white rump extending up the back. Its call is a far-carrying, rising fluty whistle *coour-lii*, while during migration it often gives a repeated *cue-cue-cew*.

Where to see Common breeder from southern Finland to southern Lapland. Found in fields, grasslands and shallow waters.

Bar-tailed Godwit *Limosa lapponica* 37cm

Large and chunky, about the size of Whimbrel. The adult male in breeding plumage has plain orange-red underparts and a very long, thick, slightly upturned dark grey bill. The tail is white with fine black barring, and a wedge-shaped white rump extends well up the back. It lacks the white wing-bars of Black-tailed Godwit. Females and juveniles are browner, with a pale eyebrow, pinkish bill with a dark tip and long dark legs.

Where to see Fairly common summer visitor. Breeds in Lapland's swamps and highlands. Migrates in large spring flocks, especially along the coast; rare in autumn.

♂ br.

juv.

Black-tailed Godwit *Limosa limosa* 40cm

Larger than Bar-tailed Godwit, but slimmer and with longer legs. Adult male has an orange head and chest, lighter than Bar-tailed, and a long, straight orange bill with a dark tip. Belly is white with dark stripes; strong eyebrow stripe, dark wings with prominent wing-bars, white upper back and black tail. Very vocal.

Where to see Uncommon summer visitor. Locally common breeder in the Oulu region, central Finland and southern Lapland. Breeds in wetlands, coastal shores and swampy areas; does not form large migrating flocks.

♂ br.

Jack Snipe *Lymnocryptes minimus* 19cm

Smaller than Common Snipe, with a short bill. Bright yellow stripes on the head and back distinguish it from other snipes. Overall very dark and cryptic, making it difficult to spot on the ground. The chest is heavily streaked, the belly white and the short tail pointed without white. Its call is a rapid churring trill, *kollorap, kollorap, kollorap,* often heard during display flights at night.

Where to see Uncommon breeder in Lapland swamps; more frequently observed during migration, and occasionally in small ditches or creeks in winter.

Woodcock *Scolopax rusticola* 35cm

Larger than snipes, chubby, with a long, straight bill. Upperparts are reddish-brown, underparts buff-coloured and the head is barred with black. In flight, it appears clumsy, flapping its short, broad wings rapidly. Its call is a soft *wart-wart-wart pissp!,* often given during dusk and dawn flights along repeated routes.

Where to see Common summer visitor, typically in woodlands. Most active at dawn and dusk, occasionally foraging near roadsides in the early morning.

Great Snipe *Gallinago media* 28cm

Similar to Common Snipe but more robust with a shorter bill. Distinguished by a heavily barred belly and prominent white outer tail feathers, especially noticeable in flight. It has a broader body and a slower, more direct flight compared to Common. Its call is a soft churring trill, often given during display flights in spring.

Where to see Rare breeder in eastern Finland, more commonly an autumn visitor. Prefers damp meadows and marshes, where it feeds and displays; difficult to observe due to secretive behaviour.

Common Snipe *Gallinago gallinago* 25cm

Size of a thrush, with a long bill, short legs and a compact, chunky body. Bold stripes mark the head and back. Unlike Great Snipe, it has a white, unbarred belly and a tail without white outer edges. Most distinctive is the mechanical courtship display, the drumming dive, which produces a vibrating tail sound, *hu-hu-hu-hu-hu-hu*. Its vocal song, a repeated *cjip-per cjip-per cjip-per,* is also commonly heard.

Where to see Common summer visitor, often flying or perched on trees and poles near wetlands, lakes or swamps.

Red Phalarope *Phalaropus fulicarius* 22cm

Larger and stockier than Red-necked Phalarope, with a shorter, thicker yellow bill tipped dark. In breeding plumage, the female is bright rusty-red below with a dark head, white cheeks and grey back; the male is similar but duller. Juveniles and non-breeding birds are mainly grey and white, with a dark cap and white face.

Where to see A rare vagrant and scarce summer visitor, usually observed at sea or occasionally on large inland lakes.

♀ br.

Red-necked Phalarope *Phalaropus lobatus* 18cm

A small, delicate wader often seen swimming in open water. The female is brighter, with a grey back, reddish-brown neck, white throat and fine white spot above the eye. The bill is thin, straight and dark. Males are similar but duller overall.

Where to see An uncommon summer visitor. Found on lakes and coastal waters during migration, often spinning in circles while feeding. Breeds on small tundra ponds and lakes in Lapland.

♀ br.

♂ br.

Marsh Sandpiper *Tringa stagnatilis* 24cm

A slender, delicate wader with a thin, straight, needle-like bill and long yellowish-olive legs. The back is pale grey-brown with faint dark spotting, the head and chest are finely streaked and the belly is white. In flight, a long, narrow, wedge-shaped rump is visible. Usually silent in Finland.

Where to see An uncommon visitor and rare breeder, mainly in southern and central Finland on marshes, shallow ponds, wet meadows and mudflats.

Greenshank *Tringa nebularia* 32cm

juv.

Larger than other *Tringa* species, with a long, slightly upcurved bill. Plumage is pale grey with a pale brown back, streaked head and chest, and a long, broad white wedge on the rump. Bill and legs are olive-green. The call is a loud, whistling *tyew-tyew-tyew*.

Where to see A common breeder and migrant throughout Finland, mainly outside the southwest. Found in shallow wetlands, marshes and flooded meadows, actively wading and feeding during both spring and autumn migration.

Terek Sandpiper *Xenus cinereus* 24cm

Smaller than Greenshank, with a long, slightly upturned bill and a steep forehead giving a distinctive profile. Plumage is pale grey above with fine dark streaks and white underparts; legs are yellowish-orange. In flight, the white trailing edge of the wings stands out. On the ground, it often runs quickly with a forward-leaning posture.

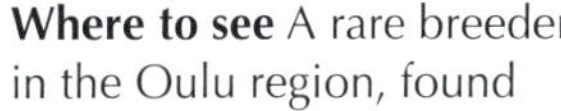

Where to see A rare breeder in the Oulu region, found on sandy shores, mudflats and shallow coastal areas.

Common Sandpiper *Actitis hypoleucos* 19cm

Smaller than Green and Wood Sandpipers, with a light brown back, white belly and distinct white shoulder patch forming a notch at the chest. It has a short bill and a thin pale eye-stripe. It constantly flicks its tail while walking. In flight, broad white wing-bars are visible. The call is a rapid series of high-pitched whistles, *swee-swee-swee-swee*.

Where to see A common breeder throughout Finland, along lake shores, rivers and beaches.

Green Sandpiper *Tringa ochropus* 22cm

Very similar to Wood Sandpiper but darker overall, with small white pearl-like spots on the back, a shorter eye-stripe ending before the eye and a sharp contrast between the streaked chest and white belly. In flight, the dark upperparts and white rump with broad black tail bars are distinctive. The call is a sharp, ringing *tuit-tuit-tuit.*

Where to see A fairly common breeder across Finland, inhabiting forest ponds, swamps and marshes, often seen perched in trees near water.

Wood Sandpiper *Tringa glareola* 20cm

Smaller than Greenshank, with a shorter bill. Very similar to Green Sandpiper but with a longer eye-stripe, prominent white spotting on the back, a diffuse pattern on the chest and yellowish-green legs. In flight shows white rump and narrow blackish bars on tail. Song is a rapid burst of yodeling: *tue-luell-tue-luell-tue-luell-tue-luell.*

Where to see Common breeder throughout Finland. Not very selective about habitat, visiting a variety of shallow water at beaches, mudflats, marshes and shores. Often actively forages at the water's edge, probing for invertebrates.

Common Redshank *Tringa totanus* 26cm

A medium-sized wader, smaller than Greenshank, with a shorter, straighter bill. The legs are bright red, and the bill is reddish at the base with a dark tip. Upperparts are brownish with fine streaks on the breast, a thin eye-ring and a short pale supercilium. In flight, the broad white trailing edges to the wings and a white wedge on the back are distinctive. The call is a sharp, ringing *teu-teu,* and the song a musical *tuel-tuel-tuel-tuel-tuelii-tuellii.*

Where to see A common summer visitor, breeding mainly along coasts, archipelagos and islands; scarcer inland.

Spotted Redshank *Tringa erythropus* 30cm

Similar to Common Redshank but larger and slimmer, giving it a more elegant appearance. In breeding plumage, it is almost entirely black, making it unmistakable. The bill is long and straight with a reddish base, and the legs are dark red to black. White spots mark the back, and in flight a long, oblong white wedge on the back is distinctive. Juveniles and non-breeding adults are grey with red legs and a long, straight dark bill. The call is a clear, disyllabic whistle, *chu-it!*

Where to see Uncommon breeder in northern Finland; migrates through coasts, bays and shallow wetlands.

Ruddy Turnstone *Arenaria interpres* 23cm

A medium-sized wader, unmistakable in breeding plumage. Adults have a striking black-and-white head, black breast, white belly, short black wedge-shaped bill, bright red legs and an orange and black back. Juveniles are browner with paler underparts and a less contrasting head but retain the characteristic bill. In flight, the rapid wingbeats, dark upperparts, and white belly and rump are distinctive.

juv.

Where to see An uncommon breeder along Finland's coasts, archipelagos and islands; rare inland. Often flips stones and seaweed on rocky shores and tidal flats while searching for invertebrates.

br.

Red Knot *Calidris canutus* 25cm

Slightly larger than Curlew Sandpiper, with a short, sturdy bill, chunky body and relatively short legs. In breeding plumage, adults have rusty-red underparts and a mottled back patterned with orange, black and grey. Plumage of adults outside the breeding season, as well as juveniles, is much plainer: grey upperparts, white underparts and a distinct white supercilium above the eye. This seasonal change makes the species variable in appearance, but its compact shape and short bill remain distinctive.

Where to see A passage migrant, mainly along the west coast, archipelagos and in southeastern Finland during migration to and from the Arctic.

br.

juv.

Ruff *Calidris pugnax* 31cm

A large wader, similar in size to Greenshank. The male is unmistakable in spring display plumage, with a raisable crest and a striking ruff showing nearly every colour variation. Legs are long and bright orange, with a medium-length, down-curved bill. Females, non-breeding males and juveniles are pale brown and much less conspicuous. Often silent.

♂ br.

Where to see Locally common breeder, renowned for its impressive spring displays. Found in the highlands of Lapland and from the Oulu region southward. Uncommon further inland. Prefers swampy areas, wet meadows and coastal marshes for feeding and breeding.

♀ ♂

Broad-billed Sandpiper *Calidris falcinellus* 17cm

Slightly smaller than a Dunlin with a long, broad, slightly downturned bill. The head shows a pale supercilium and a dark central crown stripe. In breeding plumage, the upperparts are patterned dark grey and black, the underparts are white and the breast has blackish streaks.

Where to see Uncommon in spring migration, mostly along coastal shores; rare in autumn. Breeds in swampy areas in Lapland.

Temminck's Stint *Calidris temminckii* 14cm

A very small wader, similar in size to Little Stint but appearing more elongated due to its longer tail. Plumage is plain light brown on the back, head and breast, with whitish underparts and dull yellowish legs. In flight, narrow white edges on the tail are distinctive.

Where to see An uncommon breeder in Lapland's tundra and highland wetlands. Regular during spring and autumn migration along coastal shores and mudflats, often among other small waders.

Curlew Sandpiper *Calidris ferruginea* 20cm

sum.

juv.

Slightly larger than a Dunlin, with a longer, distinctly down-curved black bill and long black legs. Breeding males are rusty-red below with dark upperparts; in flight, the white rump and narrow wing-stripes are distinctive. Juveniles are pale brown with a clear eyebrow stripe and a long, slightly down-curved bill.

Where to see Passage migrant, most often observed during spring and autumn along coastal shores and mudflats, usually among other *Calidris* waders.

♂ br.

Sanderling *Calidris alba* 20cm

br.

Similar in size to a Dunlin but more compact, recalling the shape of a small Red Knot. In spring, adults have a greyish head and breast, white belly and mottled back. In summer, the head and breast turn rufous, while the back becomes greyer. Juveniles are paler with a neat, scaly pattern on the back.

Where to see A passage migrant in Finland, most often found on coastal sandy beaches and mudflats among other small waders.

Dunlin *Calidris alpina* 19cm

Small wader, noticeably smaller than Curlew Sandpiper. Breeding plumage shows a rusty-brown head and back, distinctive black belly patch, medium-length down-curved black bill and black legs. Breast is lightly streaked. Juveniles are paler, lacking the solid belly patch, and show a neat V-shaped pattern on the back. In flight, a white wing-bar is visible.

br.

Where to see Uncommon breeder. Migrates through Finland in spring and autumn, often in flocks on coastal mudflats. Breeds sparsely on Baltic Sea islands (ssp. *schinzii*) and Lapland bogs (ssp. *alpina*).

juv.

Purple Sandpiper *Calidris maritima* 21cm

Medium-sized, stocky wader with short yellowish legs and a slightly down-curved, dark-tipped bill. Plumage is generally dark greyish-brown, sometimes with a subtle purplish sheen in good light, and underparts are pale. In flight, it shows narrow white wing-bars and dark upperparts. Blends well with rocky shorelines due to its subdued colouration.

win.

Where to see Winter visitor in Finland, mainly along rocky seashores on the west and south coasts. Rarely seen inland. Some pairs may breed in the highlands of Lapland.

Pectoral Sandpiper *Calidris melanotos* 21cm

Slightly larger than Dunlin, with a relatively short, slightly down-curved bill and olive-brown legs. A distinct white eyebrow contrasts with the heavily streaked breast, which ends sharply at the clean white belly, giving the species its characteristic pectoral-banded look.

Where to see A rare visitor to Finland, recorded only a few times each year. Typically found on mudflats, wet meadows or shorelines, often feeding among other small waders.

Little Stint *Calidris minuta* 14cm

One of the smallest waders, similar in size to Temminck's Stint. In breeding plumage it shows a rusty-orange head and back, a short straight dark bill, dark legs and a strikingly white chest and belly. Juveniles are paler, lacking the strong orange tones but show a distinctive white V-shaped mark on the back that is also visible in flight.

Where to see An uncommon migrant, regularly observed during spring and autumn along coastal mudflats, shallow beaches and occasionally inland wetlands.

br.

Arctic Skua *Stercorarius parasiticus* 40cm, wingspan 113cm

A gull-like seabird with pointed wings and fast, falcon-like flight. The dark morph is uniformly dark brown with pale primary patches and pointed central tail feathers. The pale morph shows a dark head and wings, a greyish breast and a small white patch at the base of the bill. Juveniles are more rusty-brown than the greyish Long-tailed Skua and have clear white primary patches. Flight is agile and powerful as birds chase gulls and terns.

Where to see A summer visitor. Breeds sparsely along coasts, in archipelagos and on offshore islets.

Long-tailed Skua *Stercorarius longicaudus* 38cm, wingspan 109cm

sum.

sum.

Slightly smaller and more graceful than Arctic Skua. Adult has long, slender tail feathers, a pale grey back and black cap, and it lacks the pale primary patches seen in Arctic. Flight is light, buoyant and almost tern-like. Juveniles vary from dark to greyish, with longer, narrower wings, a more pointed tail and fine barring under the tail.

juv.

Where to see Breeds in the highlands and tundra of northern Lapland. Occasionally seen on migration farther south.

Pomarine Skua *Stercorarius pomarinus* 46cm, wingspan 120cm

Larger and bulkier than Arctic Skua, with long, broad wings and a pot-bellied shape. Adults have a strong bill, large white primary patches and a distinctive spoon-shaped tail. Dark morphs resemble Arctic but show double white patches under the wings; pale morphs have a dark cap, breast-band and dusky underparts. Flight is powerful and slower than other skuas.

pale

Where to see Uncommon migrant, mainly along the Baltic Sea coast and outer archipelago in spring and autumn.

Black Guillemot *Cepphus grylle* 35cm

Smaller and more delicate than Razorbill, with glossy black plumage, a slender black bill, bright red legs and large white wing patches. Flies low over the sea with rapid wingbeats, appearing compact and short-necked. In winter, plumage turns mottled black and white but the legs remain red.

Where to see Common along the entire Finnish coast and archipelago. Nests in colonies among coastal boulders. Mainly a summer visitor, though some remain offshore through winter.

Razorbill *Alca torda* 40cm

Larger than Black Guillemot, with black upperparts, white underparts and a short, deep bill marked by a vertical white stripe. A fine white line runs from the bill to the eye, and a narrow white trailing edge to the wing is visible in flight. Flies low over the sea with rapid wingbeats, showing entirely dark upperwings.

Where to see Breeds colonially on offshore islets, especially in the outer archipelago and Åland. Mainly a summer visitor, a few overwinter.

Little Auk *Alle alle* 20cm

Half the size of other auk species, very small and compact with short wings and a short, thick bill. Plumage is black above and white below. In winter, the cheek becomes white, extending behind the eye to the crown. Flies low over the water with very rapid wingbeats.

Where to see Rare in Finland, mostly winter vagrant. Usually seen as single individuals, occasionally found inland on lakes rather than at sea.

win.

Common Guillemot *Uria aalge* 40cm

Similar in size to Razorbill but with a longer, pointed bill and dark brown upperparts. Often stands upright on rocks and cliffs. Flies low over the water with rapid wingbeats, showing mottled underwings. In winter, the face turns white with a dark line through the eye.

Where to see Uncommon breeder in the Baltic Sea, mainly in Åland and the eastern Gulf of Finland, nesting among Razorbill colonies.

sum.

Little Tern *Sternula albifrons* 23cm, wingspan 44cm

The smallest tern, nearly half the size of Common Tern. It has a white forehead, black cap and yellow bill with a black tip. The outer 2–3 primaries are dark, giving the wings a contrasting appearance. Flies with quick, fluttering wingbeats and often hovers low over the water.

Where to see Summer visitor. Small breeding population along rocky seashores near Oulu; occasionally seen at lakes and coastal areas during migration.

Sandwich Tern *Thalasseus sandvicensis* 38cm, wingspan 91cm

Medium-sized tern with a long, narrow black bill tipped in yellow. The head is black with a small crest, and the outer primaries are slightly darker, noticeable in flight. Wings are long and broad, with slower, graceful wingbeats compared to Common or Arctic Terns. Flight call is loud and distinctive, a repeated *kerrick*.

Where to see Uncommon visitor in Finland, occasionally observed along southern beaches or migrating past islands in the Baltic Sea.

White-winged Tern *Chlidonias leucopterus* 23cm, wingspan 53cm

Similar in shape to Black Tern but with more contrasting plumage. Body is mostly black with a dark grey back, while the upperwings and tail are white. Bill is short and black; legs are long and reddish. Outer primaries are dark grey, and underwing coverts are black. Flight is agile and often low over water or marshes, hunting for insects.

Where to see Rare vagrant in Finland, occasionally observed flying over lakes and wetlands during migration.

br.

Black Tern *Chlidonias niger* 24cm, wingspan 59cm

Small tern, nearly as small as Little Tern. Adult has a black body with uniformly grey upperwings and upper tail. Bill and legs are black. Juvenile has a dark cap extending behind the eye, a dark breast patch and grey upperwings similar to the adult. Often flies buoyantly over water surface like a swallow.

Where to see Summer visitor. Breeds at a few specific lakes. During migration, it can be seen flying over various lakes while hunting for insects.

br.

Caspian Tern *Hydroprogne caspia* 51cm, wingspan 104cm

br.

juv.

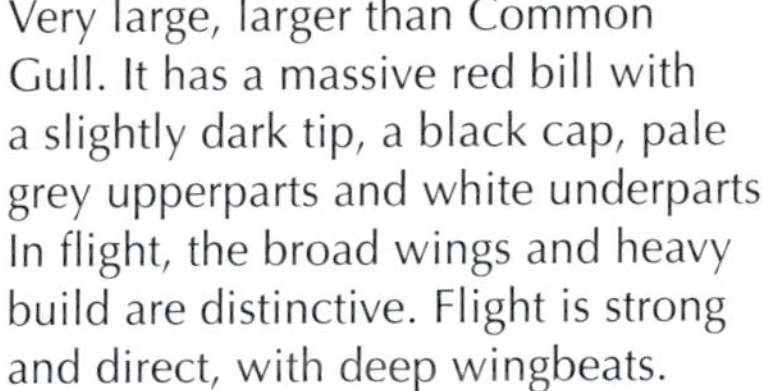

Very large, larger than Common Gull. It has a massive red bill with a slightly dark tip, a black cap, pale grey upperparts and white underparts. In flight, the broad wings and heavy build are distinctive. Flight is strong and direct, with deep wingbeats.

Its call is a loud, harsh *kraah* or *kree-aar*, carrying far over the water.

Where to see Uncommon breeder along the Baltic Sea and coastal islands; rarely seen inland on large lakes.

br.

Arctic Tern *Sterna paradisaea* 36cm, wingspan 72cm

br.

br.

Very similar to Common Tern but with a darker red bill lacking a black tip, shorter legs and a longer tail visible when perched. In flight, the long tail streamers, narrow translucent wings and smooth, agile flight distinguish it from Common Tern. Calls are similar to Common.

Where to see Summer visitor throughout Finland. Prefers the Baltic Sea coast and large inland lakes in the south, while breeding in northern Finland often occurs on remote Lapland lakes.

Common Tern *Sterna hirundo* 36cm, wingspan 75cm

br.

br.

Light grey overall with a black cap, narrow long wings, and a long, forked tail. The long red bill has a black tip, and legs are red. Flight is smooth and agile, distinguishing it from gulls. During the breeding season, it is very vocal, often calling sharp *kit-kit-kit* notes and rolling *kierri-kierri-kierri* sounds.

Where to see Common breeder on lakes and sheltered coastal bays, nesting in colonies or alone. Rare in northern Lapland.

Little Gull *Hydrocoloeus minutus* 26cm, wingspan 65cm

Smallest gull. Adult has a black head, dark underwings, small dark red bill, and bright red legs. Flies low over water with light, agile wingbeats, often hunting insects like a tern. Juveniles show dark head and upperparts, a black cap and ear-spot, black tail-band and a dark W-pattern on the wings. Underwings are mostly white with a small dark patch near the body. Very vocal during breeding season, giving rapid, short *keck* calls and *kay-ke kay-ke kay-ke* display calls.

Where to see Summer visitor. Breeds in colonies on lush lakes and coastal bays from southern Finland to southern Lapland.

juv.

Black-legged Kittiwake *Rissa tridactyla* 40cm, wingspan 99cm

win.

1st win.

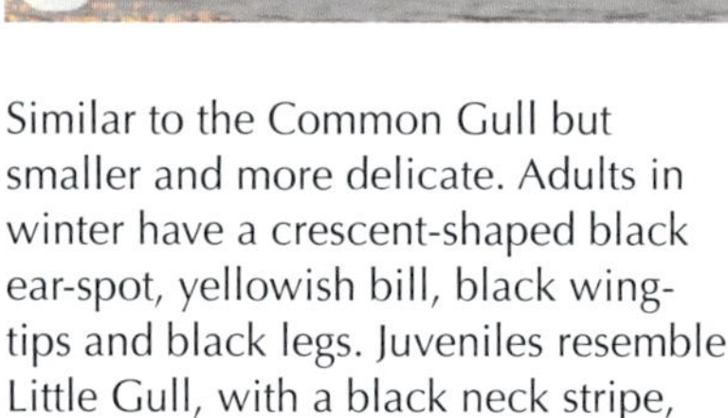

Similar to the Common Gull but smaller and more delicate. Adults in winter have a crescent-shaped black ear-spot, yellowish bill, black wing-tips and black legs. Juveniles resemble Little Gull, with a black neck stripe, light grey back, dark W-pattern across the wings, black bill, black ear-spot and black tail-band.

Where to see An uncommon visitor in Finland, mostly during autumn and winter. Found on large inland lakes and along the sea coast.

Black-headed Gull *Chroicocephalus ridibundus* 37cm, wingspan 93cm

br.

win.

Larger than Little Gull. In breeding plumage, adults have a chocolate-brown head with a white eye-ring, a dark red bill and dark red legs. In flight, shows a white leading edge on the outer wing and black primaries below. In winter, lacks the brown hood but retains the red bill and legs. Very vocal in the breeding season, calling *krreearr*.

Where to see Common breeder from southern Finland to southern Lapland, on reedy lakes and coastal bays.

Common Gull *Larus canus* 43cm, wingspan 107cm

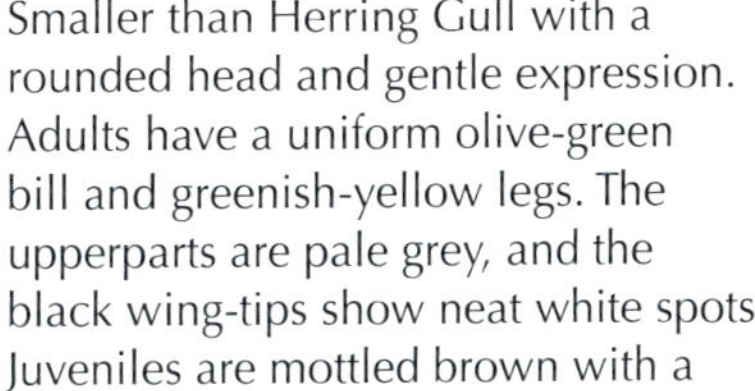

Smaller than Herring Gull with a rounded head and gentle expression. Adults have a uniform olive-green bill and greenish-yellow legs. The upperparts are pale grey, and the black wing-tips show neat white spots. Juveniles are mottled brown with a dark bill and pale face, turning greyer with age. Flight is light and buoyant, and the call is soft and mewing, less harsh than that of Herring.

Where to see Common breeder on lakes, coasts and marshes throughout Finland.

Caspian Gull *Larus cachinnans* 58cm, wingspan 142cm

Resembles Herring Gull but is slimmer, with a smaller head, longer legs and wings, and a flatter forehead. The bill is longer, thinner and more parallel-sided than Herring. Adults are pale grey above with black wing-tips; juveniles show finely patterned greater coverts and a cleaner underwing than Herring.

Where to see An uncommon visitor. In Finland, most records are of 2cy birds found at landfill sites among other gulls. Occasionally seen along coasts, on large lakes or offshore islands.

Herring Gull *Larus argentatus* 57cm, wingspan 135cm

br.

juv.

Larger than Common Gull, with a bulky build and a confident, somewhat grumpy expression. Adults have a strong yellow bill with a red spot and bright yellow legs. The upperparts are pale grey with black wing-tips showing neat white spots. Flight is powerful and steady. Juveniles are brown and heavily mottled, taking several years to reach adult plumage.

Where to see Common breeder throughout Finland, especially along coasts, large lakes and urban areas. Breeds in colonies on cliffs, rooftops and islands.

Great Black-backed Gull *Larus marinus* 68cm, wingspan 155cm

br.

br.

Largest of all gulls. Adults have a massive build with a black back, thick neck, large head, heavy yellow bill and pink legs. In flight, shows broad wings with a bold white trailing edge and more white in the wing-tips than Lesser Black-backed. Juveniles resemble juvenile Herring Gull but are bulkier with a heavier bill and larger head.

Where to see Resident and partial migrant along the Baltic Sea coast. Occasionally seen inland on large lakes, often among other large gulls.

Lesser Black-backed Gull *Larus fuscus* 52cm, wingspan 126cm

Smaller and slimmer than Great Black-backed Gull, with a more elegant build and narrower wings. Adults have dark slate-grey to black upperparts, yellow legs and a strong yellow bill with a red spot. Wing-tips are black with small white spots, less extensive than in Great Black-backed. Flight is buoyant and graceful. Juveniles are mottled brown, darkening as they age.

Where to see Summer visitor, breeding on Baltic coastal islands and some inland lakes. Locally common from southern Finland to southern Lapland.

Glaucous Gull *Larus hyperboreus* 65cm, wingspan 148cm

Very large, bulky gull with a longish bill and pale appearance. In all plumages, the whitish primaries distinguish it from Herring Gull. Adults have pale grey upperparts, white body, yellow bill with a red spot and pink legs. In winter, the head is lightly streaked brown. Juveniles and immatures are creamy-brown overall with a pink bill tipped black.

1st win.

Where to see Uncommon winter visitor, usually along the coast or on large lakes, often among flocks of Herring and other large gulls.

Iceland Gull *Larus glaucoides* 56cm, wingspan 131cm

Medium-sized, pale gull slightly smaller and slimmer than Herring Gull. Adults have pale grey upperparts, clean white underparts, pink legs and a yellowish bill with a red spot. Wing-tips are pale, lacking black markings, giving a ghostly look in flight. Juveniles are pale brown and buff-toned with whitish wing-tips and a gentle expression. Flight is smooth and buoyant.

1st win.

Where to see Rare winter visitor, mostly along coasts and large lakes. Often found in mixed flocks with Herring and other large gulls.

Red-throated Diver *Gavia stellata* 60cm, wingspan 100cm

Smaller and slimmer than Black-throated Diver. In breeding plumage, it has a pale grey head, rusty-red throat patch, red eye and a long, sharp black bill that is often held slightly upward. The neck shows fine black-and-white stripes extending to the nape. In flight, appears slender with quick wingbeats. The call is a distinctive cackling or wailing sound.

Where to see Summer visitor, breeding sparsely across Finland on small, remote lakes and bog pools. Frequently seen fishing on larger lakes or migrating with Black-throated Divers in spring and autumn.

Black-throated Diver *Gavia arctica* 70cm, wingspan 110cm

Larger and heavier than Red-throated Diver, with a thick neck and long, sharp black bill. In breeding plumage, the head is pale grey, and the dark throat often shows a purplish sheen in good light. The breast and neck have striking black-and-white vertical stripes, and the back features a distinctive white checkerboard pattern.

The haunting, far-carrying song, *clooee-co-clooee-co*, is unmistakable.

Where to see Summer visitor, breeding widely across Finland on large, open lakes. Forms flocks after breeding and migrates in groups during spring and autumn.

Great Northern Diver *Gavia immer* 80cm, wingspan 135cm

Very large diver, similar to White-billed Diver. In breeding plumage, the head and long, sharp bill are black, and the neck appears dark green in good light with black-and-white vertical stripes. A small black-and-white throat stripe is visible. The back is black with white checkerboard-like spots, similar to Black-throated Diver, and the wings show small white spots.

Where to see Rare visitor in Finland. Occasionally recorded on large lakes or along the Baltic Sea coast. Also observed during spring Arctic migration; very rare in autumn.

White-billed Diver *Gavia adamsii* 85cm, wingspan 142cm

The largest diver. In breeding plumage, the head is black, and the bill is bright yellow, almost whitish. The back is black with white checkerboard-like spots, and the neck shows black-and-white vertical stripes, similar to other divers.

Where to see Uncommon visitor in Finland. Migrates along the coast during spring Arctic migration, often in the same flocks as Black-throated. In autumn, it is very rare, occasionally seen along the Baltic Sea coast or on large, clear-water lakes.

Black Stork *Ciconia nigra* 98cm, wingspan 190cm

Slightly smaller and slimmer than White Stork. Adults have glossy black-green plumage with a striking white belly, a long straight neck, a bright red bill and legs, and bare facial skin. Juveniles are duller brown with a greyish bill and legs. In flight, the white belly and underwings contrast sharply with the dark body and long, broad wings.

Where to see A rare vagrant in southern Finland, usually observed during spring or summer, often seen soaring or feeding in open fields.

White Stork *Ciconia ciconia* 103cm, wingspan 200cm

A large, unmistakable bird with black-and-white plumage. Its long neck, long pinkish legs, and bright orange-red bill give it an elegant appearance. In flight, black flight feathers contrast sharply with the white body and upperwings. Often walks gracefully while foraging for insects, frogs and small mammals on open ground.

Where to see An uncommon visitor to Finland, mostly in the south during spring migration, occasionally wandering farther north. Usually seen in fields or wetlands.

Cormorant *Phalacrocorax carbo* 86cm, wingspan 135cm

Large waterbird with a long neck and tail. Finnish birds belong to the *sinensis* subspecies, which has a distinct whitish head and neck when breeding. The rest of the plumage is glossy black with a greenish sheen and a white patch on the thighs. The bill is thick, pale and hooked. Often seen perched with wings spread to dry.

Where to see Common breeder along the Baltic coast and on large inland lakes, nesting in colonies on islands.

Bittern *Botaurus stellaris* 75cm, wingspan 115cm

A distinctive, chunky heron-like bird with brown, heavily streaked plumage. Smaller than a Cormorant, with a short neck giving it a compact appearance. Its body is plump, its legs are relatively short and bill long and pointed. In flight, broad wings and a rounded silhouette may resemble an Eagle Owl at dusk. The deep, far-carrying *whoooomp* call is unmistakable in spring.

Where to see Summer visitor. Secretive, inhabiting reedbeds and wetlands near lakes, rivers and coastal lagoons. Occasionally overwinters in mild conditions.

Great Egret *Ardea alba* 93cm, wingspan 155cm

br.

non-br.

Similar in size and shape to Grey Heron but entirely white. Has a long, sturdy, dagger-like bill, dark during the breeding season and yellow outside it. The long neck forms a distinct S-curve in flight, which is characterised by slow, deep wingbeats on broad, rounded wings.

Where to see Scarce breeder in southern Finland, often nesting among Grey Herons. Frequently seen during migration across the country, sometimes in small flocks. Usually wades gracefully in shallow lakes or coastal bays.

Grey Heron *Ardea cinerea* 93cm, wingspan 165cm

non-br.

Large, long-legged, long-necked wading bird with grey upperparts, white neck, and black crown and plumes. Bill is long, thick and yellowish, becoming orange in the breeding season. Flies with slow, deep wingbeats, with the neck retracted into an S-shape, on broad grey wings with black flight feathers.

Where to see Common breeder across Finland, especially in southern and central regions.

br.

Found along lakes, rivers and coasts. Often stands motionless while hunting fish or frogs in shallow water.

Nightjar *Caprimulgus europaeus* 26cm

A long, slender bird with dark brownish-grey, mottled plumage that provides excellent camouflage. Has a tiny bill, large dark eyes and long, pointed wings with white patches near the tips (in males). Tail is long and slightly rounded. Call is a distinctive, dry, continuous churring sound.

Where to see Summer visitor in southern Finland. Best observed at dusk or night in dry pine forests, clearings or along sandy roads while hunting insects. Sometimes roosts motionless on branches.

Common Swift *Apus apus* 18cm

A fast, sickle-winged flier, almost entirely dark sooty brown but appearing black against the sky. Shows a small pale throat patch. Wings are long and curved, and tail is short and slightly forked. Never perches on wires or branches, clinging only to vertical surfaces. Flight is rapid and agile, with high, piercing *screee* calls often given by excited groups.

Where to see Common summer visitor throughout Finland. Breeds in buildings, including church towers, and other cavities. Almost always seen in flight, often screaming over towns and villages.

Tengmalm's Owl *Aegolius funereus* 25cm

juv.

Medium-small owl with a large, flat-topped head and compact body. Brown upperparts are dotted with small white spots, while the face is pale grey with distinct white eyebrows and bright yellow eyes. In flight, wings appear short and rounded. The male's song is a clear, rhythmic *pu-pu-pu,* repeated evenly through calm nights in spring.

Where to see Uncommon breeder, most numerous in central and eastern Finland. Prefers mature coniferous forests. Nests in old Black Woodpecker cavities or nest boxes. Strictly nocturnal, though sometimes seen roosting quietly in dense spruce by day.

Hawk Owl *Surnia ulula* 39cm

Medium-sized grey owl with a long tail and pointed wings, giving it a hawk-like look in flight. The rounded head has bright yellow eyes and a white facial disk bordered by a bold black frame. Underparts are white with dark horizontal streaks, and the back is dark brown with white spots and white shoulder patches. Flight is fast and direct, resembling a small raptor.

Where to see Uncommon breeder from the Oulu region northward to Lapland. Active mainly by day and often seen perched on treetops or poles scanning for prey. In winter, it may wander south and appear across Finland.

juv.

Pygmy Owl *Glaucidium passerinum* 17cm

The smallest owl in Europe, smaller in size than a starling. It has a rounded head, compact body and short tail, giving it a ball-like appearance. The upperparts are brown with fine white spots, while the underparts are whitish with brown streaks. Bright yellow eyes and bold white eyebrows create an alert expression. Flight is fast and darting, reminiscent of a woodpecker. The call is a clear, repeated whistle, sometimes in a rising series.

Where to see Resident. Mostly breeds in southern and central Finland, rare north of Oulu. Prefers mature coniferous forests, nesting in old woodpecker holes or nest boxes. Often seen perched high in spruce tops; in winter, may visit bird feeders to ambush small birds.

Long-eared Owl *Asio otus* 34cm

Medium-sized owl with long, narrow ear-tufts, which are usually raised when alert. Plumage is russet-brown with dark streaks, blending perfectly with tree bark. The pale face with bright red-orange eyes gives an intense look. Flight is silent and buoyant as it hunts over meadows and fields on summer nights.

Where to see Common in southern Finland, scarcer further north. Prefers mixed landscapes with forests, farmland and open fields. Roosts in dense conifers by day and hunts voles and other small mammals at night.

Short-eared Owl *Asio flammeus* 37cm

Medium-sized brown owl, similar to Long-eared Owl. It has bright yellow eyes and a small, rounded head with tiny, often invisible ear-tufts. Pale face with black eye-rings and white inner facial disc. Upperparts coarsely patterned, and tail boldly barred. Flight is smooth with slow, deliberate wingbeats, showing a pale belly, dark wing-tips, and broad white trailing edges on the upperwings.

Where to see Summer visitor from central Finland to Lapland. Often flies low over fields at night, perches on poles; it is more diurnal than most other owls. Seen throughout Finland during migration.

Snowy Owl *Bubo scandiacus* 59cm

♂

♀

Unmistakable large owl. Males are pure white, females white with narrow, dark barring on the belly and back. Rounded head with bright yellow eyes. Wings are broad and pointed, enabling powerful, direct flight.

Where to see Rare throughout Finland. Numbers depend on the abundance of lemming and vole prey. In prey scarce years, owls may migrate far south or to Siberia. In good years, a few pairs breed in Lapland mountains and occasionally individuals wander even further south, where they are seen on open fields, tundra or coastal archipelagos.

Eagle Owl *Bubo bubo* 66cm

Largest owl in Europe, with a massive head and prominent ear-tufts. Plumage is brown and heavily streaked, providing excellent camouflage in rocky and forested landscapes. Bright red-orange eyes give a fierce expression. Flight is powerful and direct, with long, broad wings built for strength rather than speed.

Where to see Resident year-round. Breeds on rocky cliffs, in quarries and occasionally in urban buildings or on the islands of the archipelagos. Hunts a wide range of prey, from small mammals to birds, using stealth and strength.

Tawny Owl *Strix aluco* 40cm

brown

grey

Medium-sized, broad-headed owl with no ear-tufts. Occurs in two main colour forms: warm brown or cooler grey. Plumage is streaked and mottled, giving excellent camouflage against tree bark. Rounded wings and deep, hooting call make it one of the most familiar owls in Europe. Large dark eyes add to its characteristic solemn look.

Where to see Widespread resident across southern and central Finland, rarer further north. Prefers mixed and deciduous forests, parks and even gardens. Nests in tree cavities, nest boxes or old nests of other large birds. Often heard more easily than seen.

juv.

Ural Owl *Strix uralensis* 55cm

Large, pale-grey owl, smaller than Great Grey Owl. It has a round head without ear-tufts, a pale facial disk and striking black eyes. Underparts are streaked, and the long tail gives it a slim, elongated profile in flight. Unlike some owls, it lacks pale wing flashes. Chicks are fluffy grey balls. Powerful and fearless when defending its nest, it may attack intruders.

Where to see Widespread across Finland, especially in forested areas. Breeds in mature pine and mixed forests, nesting in tree cavities or large nest boxes. Most active at dusk and night but often visible by day.

juv.

Great Grey Owl *Strix nebulosa* 64cm

Majestic and very large, one of the world's biggest owls. Broad, rounded head gives a striking front view, while the side profile appears flat-cut. Facial disk is pale grey with bold black streaks, concentric white whiskers and a dark patch around the yellow bill. Despite its size, it is lightweight and silent in flight, relying on the length of its wings rather than their strength.

Where to see Uncommon breeder. Prefers extensive coniferous and mixed forests near open bogs and clearings. Nests on broken stumps, old raptor nests or specially placed platforms. Most often seen perched prominently, scanning for voles, mainly in central and northern Finland.

Osprey *Pandion haliaetus* 56cm, wingspan 167cm

A large, long-winged raptor with white underparts and dark brown upperparts. The head is white with a broad dark eye-stripe. Wings are long and angled, pale below with a strong dark carpal patch and dark stripe. Tail is grey-brown with broad bars. Flight is powerful with deep, elastic wingbeats, often hovering before plunging feet-first into water. Juveniles resemble adults but show buff fringes on upperparts and more breast streaking.

Where to see Widespread summer visitor in Finland. Breeds near lakes and rivers. Frequently observed fishing or perched on dead trees, posts and pylons.

Honey-buzzard *Pernis apivorus* 55cm, wingspan 142cm

A slim, long-winged raptor, resembling Common Buzzard but with a smaller head, longer neck and more protruding tail. Plumage is highly variable, ranging from pale grey-brown to dark brown. Adults show two dark wing-bars, a dark trailing edge and a long, narrow tail with several dark bands. Males have a pale grey head, while females and juveniles are browner. Flight is buoyant and elastic, with deep, flexible wingbeats.

Where to see Summer visitor breeding in mixed forests across Finland, often seen during migration.

Short-toed Eagle *Circaetus gallicus* 66cm, wingspan 180cm

A large, long-winged eagle with a broad, dark head and distinctive underwing pattern. Upperparts are brown, contrasting with whitish underparts and grey tail with broad dark bars. The head and body appear plain and unmarked, giving a clean look. Primary tips are greyish rather than black as in Common Buzzards. Juveniles are warmer-toned, with less contrast. Specialised in hunting reptiles, especially snakes, and often seen hovering before diving.

Where to see Rare vagrant, occasionally reaching southern Finland, mostly along the coast at migration watchpoints.

Lesser Spotted Eagle *Clanga pomarina* 60cm, wingspan 155cm

A medium-sized eagle, smaller and slimmer than Greater Spotted Eagle, with long wings and a relatively small head and bill. Plumage is warm brown, with slightly darker flight feathers and a small whitish patch at the base of the primaries, often visible in flight. Adults show a distinct pale spot on the upperwing coverts, while juveniles have pale spotting and a white band across the wings. Flight is steady and gliding, wings held in a shallow arch.

sub-ad.

Where to see Uncommon visitor, mostly observed on migration in southern Finland, especially along the coast at migration watchpoints.

Greater Spotted Eagle *Clanga clanga* 64cm, wingspan 170cm

A large, dark brown eagle with broad wings and a relatively small head. Plumage is uniformly dark chocolate-brown with slightly paler flight feathers and faint lighter markings on the upperwing coverts. The tail is short and rounded. Juveniles show distinct pale spots on the upperparts and a whitish comma mark near the primaries. Flight is powerful and direct, with deep wingbeats and long glides.

juv.

Where to see Uncommon visitor, occasionally observed in southern Finland, mainly at coastal migration watchpoints and in the eastern parts of the country.

Steppe Eagle *Aquila nipalensis* 68cm, wingspan 190cm

A large, broad-winged eagle with a long, slightly wedge-shaped tail and a powerful appearance. Plumage is mostly dark brown with a contrasting paler nape forming a subtle shawl. The bill base is yellow and notably large. In flight, it shows broad wings with distinct pale panels on the upperwings and a darker trailing edge. Juveniles are paler and show a white band across the underwings.

juv.

Where to see Rare visitor to Finland, mainly observed along the southern coast during migration.

Golden Eagle *Aquila chrysaetos* 87cm, wingspan 200cm

juv.

A large, powerful eagle with long, broad wings and a long, slightly wedge-shaped tail. Adults are dark brown with golden-buff feathers on the nape, giving the characteristic golden look. Juveniles are paler below with white patches at the base of the tail and underwings, which darken with age. Flight is strong and soaring, with slow, deep wingbeats and long glides over open landscapes. The call is a high, piping whistle, rarely heard at distance.

Where to see Resident and partial migrant. Breeds in northern and eastern Finland, mainly in open forests, fells and remote uplands.

Sparrowhawk *Accipiter nisus* 35cm, wingspan 70cm

A small, slim hawk with short, rounded wings and a long, square-tipped tail, built for rapid flight through woodland. The adult male is bluish-grey above with orange-barred underparts and red eyes. The larger female is brown above with fine grey barring below and yellow eyes, often mistaken for a small Goshawk. Juveniles are brown above with streaked underparts and yellow eyes. Flight is fast and agile, with bursts of quick wingbeats followed by glides. Call is a sharp *kek-kek-kek*, especially near the nest.

Where to see Resident and partial migrant. Widespread across Finland, common around forests, villages, and gardens, often surprising small birds near feeders.

Goshawk *Accipiter gentilis* 53cm, wingspan 100cm

Large, powerful hawk with broad, rounded wings and a long tail, resembling an oversized Sparrowhawk. Adults are slate-grey above with fine grey barring below, a distinct pale supercilium and fierce red eyes. Females are noticeably larger and bulkier than males. Juveniles are brown above with heavily streaked underparts and yellow eyes, gradually turning greyer with age. Flight is strong and purposeful, with deep wingbeats and bursts of explosive speed when hunting. Call is a harsh *kek-kek-kek*, especially in the breeding season.

Where to see Resident and partial migrant. Breeds throughout Finland in mature forests. Usually elusive but sometimes seen gliding above treetops or ambushing prey along forest edges.

♀

juv.

♂

Marsh Harrier *Circus aeruginosus* 49cm, wingspan 127cm

A large harrier. Broader-winged and heavier than Hen, Montagu's or Pallid Harriers, with slow, elastic flight over wetlands. An adult male has a brown body, pale head, grey wings with black tips and a grey tail. Females are dark brown overall with a golden-yellow crown and throat, creating a distinctive pale-headed appearance. Juveniles are dark brown with a buff-yellow crown and plain body, lacking the female's barred tail. First-year birds may show paler mottling. In all plumages, the long tail and broad wings give a steady, buoyant flight low over reedbeds.

Where to see Summer visitor. Breeds widely across Finland in large reedbeds by lakes, bays and wetlands. Most visible when quartering low over marshes and fields. Migrates south in autumn.

♂

Hen Harrier *Circus cyaneus* 50cm, wingspan 110cm

♂

♀

Medium-sized harrier, larger and heavier than Montagu's or Pallid Harriers, with broader wings and less buoyant flight. The adult male is pale grey above with broad black wing-tips and whitish underparts. The body and wings are similar in tone, giving a less contrasting appearance than Montagu's. Adult females are brown above with streaked underparts, a pale rump and a strongly barred tail, appearing bulkier than Pallid. Juveniles are dark brown above with rich rufous underparts, often lacking the pale collar shown by Montagu's and Pallid. All plumages display the characteristic white rump in flight.

Where to see Uncommon breeder in open bogs and wetlands. More often observed on migration, especially in autumn, when hunting low over fields and marshes.

juv.

Pallid Harrier *Circus macrourus* 45cm, wingspan 107cm

♂

Slender and graceful harrier, similar in size to Montagu's Harrier, with long, narrow wings and buoyant, elastic flight. The fifth primary is short, as in Montagu's. Adult males are very pale grey, almost whitish, with narrow black wing-tips, a narrower black wedge than Hen Harrier and a distinct white collar. The body and wings show little contrast, giving a clean, pale impression. Adult females are brown above with streaked underparts, a pale rump and a finely barred tail, appearing slimmer and more elegant than Hen. Juveniles are rufous below with less streaking, often showing a pale collar.

Where to see Rare breeder in open farmland and marshes of Oulu region and Lapland. Juveniles are regular autumn migrants, often hunting low over fields.

♀

juv.

Montagu's Harrier *Circus pygargus* 45cm, wingspan 107cm

Slim and elegant harrier, smaller and lighter than Hen Harrier, with long wings and buoyant, graceful flight. The fifth primary is short, as in Pallid Harrier. Adult males are pale grey, slightly darker on the body than the wings, with narrow black wing-tips and faint black bars on the upperwing secondaries. Underwing shows two black stripes with a brownish-red pattern. Adult females are brown above with streaked underparts, a pale rump and a barred tail. Juveniles are dark brown with uniform rufous underparts, resembling first-year birds that may show faint streaking.

Where to see Rare breeder in southern Finland, mainly in open farmland and marshes. Juveniles are more often seen during autumn migration, foraging over fields and open landscapes.

♀

♂

juv.

Red Kite *Milvus milvus* 66cm, wingspan 185cm

Large, elegant raptor with long, angled wings and a deeply forked tail, more pronounced than in Black Kite. Plumage is reddish-brown with pale streaking, contrasting with a grey head and rufous body. In flight, the long tail twists constantly, helping it turn easily. Wing-tips are black with pale patches on the inner primaries. Juveniles are duller with a streaked buff head and less vivid body tones. Flight is graceful and buoyant, with frequent glides and effortless turns.

Where to see Rare vagrant in Finland, mainly during migration. Breeds more commonly further south in Europe.

Black Kite *Milvus migrans* 53cm, wingspan 142cm

Medium-sized raptor, slimmer than a Common Buzzard, with long wings and a slightly forked tail. Plumage is mostly dark brown, with paler flight feathers that contrast a little with the darker coverts. The head and neck are often lighter, giving a hooded appearance. Juveniles are browner with streaked underparts and a pale panel in the wings. In flight, the tail twists often as the bird glides and soars easily.

Where to see Rare breeder in southern or eastern Finland. Usually near lakes, rivers and coasts.

White-tailed Eagle *Haliaeetus albicilla* 85cm, wingspan 210cm

Europe's largest eagle, with a big body, broad wings and a short, wedge-shaped tail. Adults are dark brown with a pale head and neck, and a bright white tail that contrasts with a large yellow bill. Juveniles are dark brown all over, with mottled underparts and tail, slowly turning white over several years. Flight is strong and steady, with slow wingbeats and long glides, often described as like a flying barn door. The voice is a loud, harsh yelp, especially near nests.

Where to see Common breeder at coasts, large lakes and rivers. Resident in the south, migratory in the north.

Rough-legged Buzzard *Buteo lagopus* 55cm, wingspan 137cm

Large buzzard, with longer wings and a slimmer body than Common Buzzard. The head is small and the tail long. Underparts are usually pale with a dark belly patch and dark carpal patches on the underwings, very noticeable in flight. Upperparts are brown with paler feather edges. The tail is white with a broad black terminal band. Flight is light and buoyant, often hovering like a kestrel. Juveniles resemble adults but with less distinct belly patches and paler tones. Call is a soft, plaintive mewing, less often heard than Common.

Where to see Uncommon breeder in tundra in northern Finland; autumn migrants seen over fields and coasts.

juv.

Common Buzzard *Buteo buteo* 54cm, wingspan 120cm

Medium-sized raptor with broad, rounded wings and a short, fan-shaped tail. Plumage is highly variable, ranging from very dark brown to pale cream, though most have a brown back, streaked underparts, and a pale band across the breast. The tail is grey-brown with narrow dark bars. Flight is steady, with frequent glides on slightly raised wings, often circling over open land. Juveniles are paler below with finer streaking and less distinct tail-bars. Voice is a distinctive mewing call, often heard while soaring.

Where to see Common breeder. Widespread across Finland, especially in southern regions, where frequent in forests, fields and clearings.

Hoopoe *Upupa epops* 27cm

Unmistakable medium-sized bird with a long, flexible crest. Plumage is pinkish light brown, with bold black-and-white patterns on wings and tail, and a long, slender, down-curved bill. In flight, it appears broad-winged, almost butterfly-like. The crest is usually lowered but can be raised when landing or displaying.

Where to see Uncommon visitor in Finland during spring and autumn migration, mainly in southern and central regions. Prefers dry, open areas and grass-lands, foraging on the ground by probing soil for insects and worms.

Roller *Coracias garrulus* 31cm

Robust, crow-like bird with a large head and strong black bill. Body and most of wings are bright turquoise-blue, while the back and innerwing areas are warm brown. In flight, it resembles a blue Jackdaw, with slow, powerful wingbeats and elegant twists and turns.

Where to see Rare visitor in Finland, mostly in the south. Prefers dry, open areas and fields, often perching on wires, fence posts or atop juniper bushes while scanning for prey.

Bee-eater *Merops apiaster* 27cm

Unmistakable, thrush-sized bird with a slim body, long wings and a rounded tail ending in a small spike. Large head, long down-curved bill and bright red eyes. Plumage is vivid: a yellow throat bordered in black, blue underparts and flight feathers, and orange-brown upperparts with yellow shoulders. Call is a rolling *prrrt*.

Where to see Rare spring visitor in Finland, with only a few records each year. Mostly observed in the south and on islands, often briefly in flight during migration.

Kingfisher *Alcedo atthis* 18cm

Small, plump, colourful bird. Head and wings are bright blue, back light blue, belly orange, with an orange cheek patch. There is a distinct white line on the neck and a white throat. In flight, back and tail appear vivid light blue. Call is a sharp *tsii*, often heard while flying.

Where to see Uncommon breeder in southern Finland. Almost always near rivers; in winter, some birds may move to the coast if rivers freeze.

Wryneck *Jynx torquilla* 17cm

Slightly larger than Lesser Spotted Woodpecker, the Wryneck is distinctive with an oblong body and cryptic, barred plumage. Upperparts are grey-brown with dark bands, the wings are brown, the belly white and the throat buff-brown. Strong dark bands run across the back and behind the eyes. It has a short, pointed bill and a long tail. True to its name, it moves its neck in a snake-like manner. Song is a series of loud, whining notes: *tie-tie-tie-tie*.

Where to see Summer visitor, often hard to spot. Common in southern Finland, rarer north of Oulu. Breeds in open deciduous or mixed forests, yards and parks, using old woodpecker holes.

Grey-headed Woodpecker *Picus canus* 29cm

Larger than Great Spotted and White-backed Woodpeckers, this species has a lead-grey head with black lores and a narrow black stripe from the bill base. Males show a red patch on the front of the crown. The bill is relatively short. In flight, it appears fast and straight, with a short neck and a yellow-green rump. The territorial call in spring is distinctive: sharp *kii-kii-kii* notes that gradually drop in pitch, ending with *kuue-kuue-kuue*.

Where to see Locally common in southern Finland, rarer in the north. Breeds in deciduous woodlands, parks and alongside rivers and lakes; visits feeders in winter.

♀♂

Black Woodpecker *Dryocopus martius* 43cm

A large, crow-sized woodpecker with entirely black plumage. Males have a bright red crown, while females show red only on the rear of the head. Its long, powerful bill is used for drumming, the sound of which carries over long distances. In flight, it moves straight and strong with deep wingbeats. Calls are loud and varied: a sharp, dry *kree-kree-kree* in flight, a far-carrying *kliiih* as a territorial call, and resonant drumming.

♂

Where to see Resident and fairly common throughout Finland, most abundant in the south and rare in Lapland. Nests in large cavities in mature aspens and pines, which are later used by species such as Goldeneye and Tengmalm's Owl.

♀ ♂

Three-toed Woodpecker *Picoides tridactylus* 23cm

Smaller than Great Spotted Woodpecker, with predominantly dark plumage. A black stripe runs across the white cheek, and the underparts are streaked. A broad white stripe extends down the centre of the back. Males have a dirty-yellow crown, while this is white in females. The call is similar to Great Spotted but softer. Drumming is usually performed on dead trees, producing a slow, far-carrying echo.

Where to see An uncommon breeder across Finland. Prefers old coniferous forests, particularly spruce, where it frequently peels bark from trunks in search of insects.

White-backed Woodpecker *Dendrocopos leucotos* 27cm

Slightly larger than Great Spotted Woodpecker, with a distinctive white lower back and rump. Underparts are streaked, vent is pink and the long bill is adapted for peeling bark. Males have a full red crown, but females show black only. Wings display broad white bars and spots, creating a striking checkered pattern in flight. Drumming is long and powerful, often ending abruptly. Calls are sharp, explosive *kik* notes, similar to Great Spotted.

♂

♀

Where to see Rare breeder in southeast Finland, mainly in old, rich deciduous forests with abundant dead wood, especially black alder, birch and aspen. A strong indicator of biodiversity.

Lesser Spotted Woodpecker *Dryobates minor* 15cm

A very small woodpecker, black-and-white overall with a weak bill, barred back and streaked belly. Male's crown is red, while female's is black. Undertail is unmarked. Song is a series of 8–15 soft, piping notes, slowing at the end: *piit-piit-piit,* gentler than that of Wryneck. Drumming is soft and less conspicuous than in larger woodpeckers.

♂

♀

Where to see More common in southern Finland up to the Oulu region, rarer in Lapland. Prefers groves, lush deciduous forests and woodlands with decaying trees for nesting.

Great Spotted Woodpecker *Dendrocopos major* 25cm

A medium-sized woodpecker and the most common black-and-white species. Large white oval wing patches are the clearest field mark. Underparts are plain with a bright red vent. Male shows a red nape patch, female a black crown, and juveniles a full red crown with barred white wing patches. Flight is undulating, typical of woodpeckers. Calls include a sharp *kick* and a rattling *krrr* in alarm. Drumming is fast, powerful and carries far.

Where to see Widespread resident throughout Finland, in forests, groves, parks, gardens and often at feeders in winter.

♀ ♂

Merlin *Falco columbarius* 30cm, wingspan 60cm

Finland's smallest falcon. Short, sharply pointed wings and a dark vertical cheek-stripe. Adult males have charcoal-grey upperparts, darker wings, a black tail tip, and orange-streaked underparts and neck. Females and juveniles are dark brown, more Sparrowhawk-like in appearance, with streaked underparts and a slightly bulkier build. Flight is fast and agile, often low and direct, with rapid wingbeats interspersed with glides.

Where to see Uncommon breeder in northern Finland, favouring dead trees in logged forests. During migration, it can be observed anywhere in the country, often perched on rocks, wires or along roadsides, hunting small birds and insects.

♀

juv.

♂

Red-footed Falcon *Falco vespertinus* 31cm, wingspan 70cm

Small falcon, similar in size to a Hobby. Adult males are unmistakable, with dark grey plumage, red legs and red underparts; upperwing feathers appear silvery-grey in flight. Females have a rusty-yellow crown and belly, white cheeks with a black moustachial stripe, and a blue-grey back. In flight, females show dark grey upperparts with streaked underwings and a rust-coloured belly; the tail has a black terminal band. Juveniles are brown with a white cheek, black moustachial stripe and streaked underparts.

Where to see Uncommon visitor. Spring birds appear in southern Finland over open fields or perched on wires; juveniles may arrive in large autumn irruptions.

♂

♀

juv.

Common Kestrel *Falco tinnunculus* 34cm, wingspan 74cm

About the size of a Sparrowhawk, this falcon's wings are pointed but broader than most falcons, with the tips sometimes appearing almost rounded. The back is reddish-brown with dark-tipped wings, grey head and faint moustachial stripes. Both sexes show dark spots on the back and breast. The tail is long – greyish in males, barred in females – with a dark terminal band. Underparts are paler and evenly streaked. Often hovers while hunting over open fields.

♂

juv.

Where to see Common breeder in southern Finland, scarcer in north. Nests in nest boxes, holes in dead trees and barn walls; frequently perches on poles or wires.

Hobby *Falco subbuteo* 32cm, wingspan 65cm

A small falcon, slightly larger than a Merlin, with long, sharply pointed wings. Males and females are similar: dark grey upperparts, black moustaches, white cheeks and throat, red trousers and dark-streaked underparts and underwings. Juveniles are dark brown with yellowish cheeks and belly distinguishing them from Red-footed Falcon. Flight is fast, agile and skilful, often involving impressive aerial manoeuvres.

juv.

Where to see Summer visitor and fairly common breeder throughout Finland. Frequently hawks insects in the air and perches on trees, poles and wires.

Gyrfalcon *Falco rusticolus* 58cm, wingspan 122cm

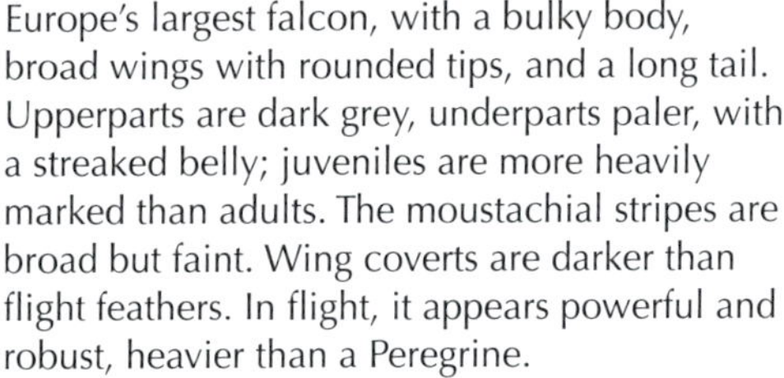

Europe's largest falcon, with a bulky body, broad wings with rounded tips, and a long tail. Upperparts are dark grey, underparts paler, with a streaked belly; juveniles are more heavily marked than adults. The moustachial stripes are broad but faint. Wing coverts are darker than flight feathers. In flight, it appears powerful and robust, heavier than a Peregrine.

Where to see Uncommon breeder on rocky cliffs in Lapland. Juveniles occasionally move south in winter, sometimes reaching central Finland.

Peregrine Falcon *Falco peregrinus* 45cm, wingspan 100cm

Large, powerful falcon with long, pointed wings and a broad tail. Adults are dark grey above with white cheeks, prominent black moustachial stripes and pale underparts with bold horizontal barring. Juveniles are browner with vertical streaks on yellowish underparts. Flight is fast and direct, with lightning-fast stoops when hunting.

Where to see Summer visitor, breeding on rocky cliffs and in open swamps in northern Finland. During migration, widely observed across the country, especially in archipelagos and coastal bays hunting waterfowl and waders.

Golden Oriole *Oriolus oriolus* 24cm

About the size of a Fieldfare. Adult males are unmistakable, bright yellow with black wings and a strong dark red bill. Females and juveniles are olive-green with a paler bill, and juveniles show streaking on the belly that fades with age. Song is a loud, flute-like whistle, *hoo-ee-hoo-ee-oo*, sometimes resembling a distant Blackbird.

♂

Where to see Uncommon breeder in southeastern Finland. Prefers deciduous forests, especially birch, aspen, and alder, often near water; favours coastal birch woodlands.

♀

Red-backed Shrike *Lanius collurio* 17cm

A small, striking shrike. Adult males have a chestnut-brown back, pale grey head with a black facial mask, pinkish belly, white throat, and a short, hooked black bill. Females and juveniles are browner, with a less distinct mask, barred underparts and a greyish nape. In flight, the long black tail with prominent white flashes is noticeable.

♂

Where to see Summer visitor, more common in southern Finland up to Oulu and rare in Lapland. Prefers open, semi-natural habitats like juniper meadows, overgrown fields, bushy logging areas and coastal meadows in the archipelago.

♀

Great Grey Shrike *Lanius excubitor* 24cm

The largest shrike, notable for its long tail and strongly patterned plumage. A light grey head and back contrast with black wings, a black facial mask and a strong, hooked black bill. In flight, it shows a small wing patch (nominate form *excubitor*) and a long black tail with white outer feathers. Call is a drawn-out, drill-like *prrriih*.

Where to see Breeds sparsely across Finland but absent from the southern coast and archipelago. Most common in northern and western areas. Prefers semi-open habitats like bogs, logging areas, sparse pine forests and mountain birch. Seen nationwide in winter.

Lesser Grey Shrike *Lanius minor* 20cm

Smaller than Great Grey Shrike, with a grey head and back, black wings, white throat, pinkish belly, long tail and a strong, hooked black bill. The black forecrown distinguishes it from Great Grey. In flight, shows larger white patches on the wings and more white in the tail than Great Grey.

Where to see An uncommon visitor in southern Finland. Prefers open fields, farmland and bushy areas, often perching on wires, poles or small trees while hunting insects and small prey.

Siberian Jay *Perisoreus infaustus* 28cm

Smaller than the Eurasian Jay and only slightly larger than a thrush. Plumage is greyish-brown with a dark brown crown, long rusty-red tail and subtle wing patch. Flies silently, gliding smoothly with rapid wingbeats. Often moves in small flocks. Generally quiet but occasionally gives harsh calls or whistles.

Where to see Resident and very local. Breeds in old-growth northern forests, rare in central Finland. More easily observed outside the breeding season, at feeders or in picnic areas where people provide food.

Eurasian Jay *Garrulus glandarius* 34cm

Plumage is reddish-grey with black moustachial stripes, pale light blue wing panels, small white wing patches, a black tail and bright white upper tail coverts. Flies on rounded wings. Calls are harsh and screeching.

Where to see Resident throughout Finland, with occasional autumn movements. Common breeder in southern spruce forests, less numerous further north. Frequently visits bird feeders in winter.

Magpie *Pica pica* 45cm (incl. 20cm tail)

Unmistakable black-and-white bird with a very long tail. The head, breast and back are black. Primaries mostly white with black edges and tips, showing a greenish metallic gloss; secondaries blue-black. The tail is long and dark, with a greenish gloss at the base and bluish-purple sheen at the tip. Underparts and shoulders are white; undertail coverts black. The voice is a mix of harsh, unmusical calls.

Where to see Resident and common throughout Finland. Frequently observed in towns, parks and residential areas, sometimes gathering in flocks.

Nutcracker *Nucifraga caryocatactes* 34cm

About the size of a Eurasian Jay but with a longer, heavier bill. Plumage is dark brown with dense white spots on the body. Wings are blackish, undertail coverts white, and a white band crosses the tail. Flight is wobbly, resembling the Eurasian Jay.

Where to see An uncommon breeder in south and central Finland. Most easily observed in autumn, when the eastern subspecies *macrorhynchos* moves west in search of seeds from Siberian pine.

Jackdaw *Coloeus monedula* 32cm

Smaller than a Hooded Crow or Rook, with a shorter bill and wings. Plumage is black and grey with pale eyes, making it easily distinguishable from other crow species. Often seen in flocks, it is very vocal, producing repeated *kja-kja-kja* calls.

Where to see Common breeder and partial migrant. Prefers urban and modified environments such as city centres and villages, from southern Finland up to the Kuusamo region. Uncommon in Lapland.

Rook *Corvus frugilegus* 45cm

Smaller than a Raven, with glossy black plumage and a grey bill. Adults have a pale, bare base to the bill. This feature is absent in juveniles, making them resemble Carrion Crows, but the size and shape of the bill helps distinguish them. In flight, it resembles a dark Hooded Crow.

Where to see A summer visitor, breeding in colonies near large fields and towns, mainly along the west coast. During migration, Rooks can be seen more widely, especially in southern Finland.

Hooded Crow *Corvus corone cornix* 48cm

Medium-sized crow with striking grey-and-black plumage. The body is pale grey, while the head, throat, wings, tail and thighs are black. Bill and legs are also black. The sharp division of colours distinguishes it from Carrion Crow and Rook. Juveniles look duller, with browner grey tones.

Where to see A common resident across Finland. Occupies towns, farmland, coasts and forest edges. An opportunistic feeder, usually seen in pairs but also gathering in small groups.

Raven *Corvus corax* 60cm

The largest crow species, with a thick, sturdy black beak, fluffy throat feathers and glossy, all-black plumage. In flight, it shows long, broad wings and a wedge-shaped tail, with slow wingbeats, steady soaring and frequent circling like a large bird of prey. Shy but sometimes curious. Calls are varied, the most common being a harsh, deep *croak*.

Where to see Resident throughout Finland. Prefers pine forests and open landscapes, and is often seen scavenging carcasses at roadside, like other crow species.

Penduline Tit *Remiz pendulinus* 10cm

Small, compact bird. Male shows a broad black mask, while the female's is narrower, running only through the eye. Head is pale grey, the back deep rufous and the chest a paler rufous. Bill is sharply pointed and conical. Very active as it moves through bushes and reeds. The call is a soft, whistled *tsiiiu*.

Where to see Uncommon breeder in southern Finland. Prefers lush bays, lakes and riverbanks with scrubby reedbeds, deciduous parks or shrub meadows. Builds a distinctive hanging bag-like nest, usually at the tip of birch, willow or alder branches.

Blue Tit *Cyanistes caeruleus* 11cm

Small, plump and fluffy. Bright yellow belly, white cheeks, green back and sky-blue cap, wings and tail. A dark blue line runs across the head, with a vertical line down the middle of the yellow belly. Narrow white wing-bars are visible. Female is slightly paler, and juveniles more yellowish with greenish-blue parts. Song is a high-pitched *siiiih-siiih si-syrrr*, with varied tit-like calls.

Where to see Common breeder up to southern Lapland, in deciduous and mixed forests, gardens, parks and yards. Partial migrant. Visits feeders in winter and readily uses nest boxes in spring.

Great Tit *Parus major* 14cm

Large, robust tit with a yellow belly marked by a strong black central line. Black head with white cheeks and a sturdy bill. Back is greenish, wings light blue with a white panel. Female is slightly paler, and the black belly line may be broken. Juveniles are pale yellowish. Vocalisations include varied calls, including *pink pink* and *tii-tuui*, and the song is a simple mechanical *ti-ti-taa*.

Where to see Common breeder throughout Finland in forests, parks and near humans. Nests in cavities or nest boxes. Partial migrant, often visiting feeders in winter.

Coal Tit *Periparus ater* 10cm

Small, plump tit with a large black cap and bib, giving it a dark appearance. White cheeks, a small white neck-patch and a lead-grey back contrast with pale brown underparts. Wings show two white wing-bars, and the tail is short.

Where to see Resident in southern Finland. Prefers coniferous forests, especially spruce, but will also use nest boxes near suitable forest. Some years, irruptive movements bring individuals farther southwest.

Crested Tit *Lophophanes cristatus* 11cm

Similar in size to Blue Tit. Its triangular, striped crest gives it a punk-like appearance. Plumage includes a brown back, pale brown underparts, a large black throat patch and a black stripe through the white cheeks, giving the face a slightly dirty look. Eyes are reddish. Call is a cheerful, distinctive trill, often described as *burrurret*.

Where to see Uncommon breeder from southern Finland to the Oulu region, rare in southern Lapland. Prefers coniferous forests, thriving in rocky pinewoods and old spruce forests. Less dependent on decaying deciduous trees than the Willow Tit.

Marsh Tit *Poecile palustris* 12cm

Very similar to Willow Tit but with a few key differences. The black cap is glossier, the black bib smaller, the back more brownish, and it lacks a pale forearm panel. Its call is distinctive and lively, often an explosive *pichay* or a longer series *zee-zee–chay-chay*, along with other calls typical of tits.

Where to see Uncommon visitor in southern Finland. Closest breeding populations are in the Baltic countries, but it occasionally wanders north. Often visits feeders in winter alongside other tits.

Willow Tit *Poecile montanus* 12m

This tit has a neat, dull black cap, a small black bib, white cheeks, a greyish back and a slightly longer tail than Coal Tit. Underparts are pale grey to whitish, and the wings show a light forearm panel. Overall plumage has more grey tones than brown. The call usually starts with two high notes followed by three or four lower notes: *zi-zi-taah-taah-taah*.

Where to see Uncommon breeder throughout Finland. Prefers older coniferous or mixed forests with suitable stumps for nesting. Excavates nest holes in soft stumps but may use natural cavities or rarely nest boxes.

Siberian Tit *Poecile cinctus* 13cm

Similar in size to Willow Tit but more colourful. White cheeks, tawny-brown back, rusty flanks, dull grey-brown cap and a larger black bib than Willow. Wings are greyish with a white panel. Call resembles Willow Tit's *zi-zi-tah-tah-tah* but harsher and shorter.

Where to see Uncommon resident from the Kuusamo region to Lapland. Prefers pine and mixed forests with fallen trees or stumps for nesting. Will also use nest boxes.

Long-tailed Tit *Aegithalos caudatus* 14cm

A tiny, plump, very active bird with a long tail making up nearly half its length. The Scandinavian race is mostly black and white with a pure white head, a small black bill, and some pinkish tones on the flanks and back. It often moves in lively family flocks, calling constantly with soft, musical *tii-tii-tii* and gentle *tsrr-tsrr* sounds.

Where to see Uncommon breeder from southern Finland to the Oulu region, rare in Lapland. Prefers mixed and deciduous forests with birch, willow and alder. Most visible in autumn and winter when flocks visit gardens and bird feeders.

Bearded Reedling *Panurus biarmicus* 15cm

♂

♀

A charming, small bird with a rounded body and a long, narrow tail. All plumages show warm brownish-orange tones, a tiny orange bill and black legs. Adult males are striking with a pale grey head, white throat and bold black moustachial stripes. Females and juveniles are paler, with a light brown head and white throat, lacking the male's facial markings. Their flight is bouncing and restless, often accompanied by a distinctive metallic *ping* call.

Where to see Uncommon resident from southern Finland to the Oulu region, breeding in reedbeds along Baltic bays. In autumn and winter, they often move in small flocks.

Woodlark *Lullula arborea* 14cm

Smaller than Skylark, it has a shorter tail and broader wings. Shows clear white wing-bars, a short crest and pale eyebrow stripes that meet on the nape, producing a gentle expression. Plumage is warm brown with fine streaks on the upperparts and paler below. The song is soft, melodic and fluting, often delivered during a long, circling flight.

Where to see Uncommon breeder in southern Finland. Prefers dry heathlands, clear-cuts and sparse pine forests. A migratory species, arriving early in spring and leaving in autumn.

Skylark *Alauda arvensis* 17cm

A medium-sized, brown and heavily streaked lark with a short tail and broad, rounded wings. Often shows a small crest that it raises when alarmed or displaying. Underparts are whitish with fine streaking on the breast. In flight, it shows white trailing edges to the wings and white outer tail feathers. The song is a long, melodious series of trills and notes, delivered high in the sky during display flight.

Where to see Common breeder throughout Finland, scarcer in northern Lapland. Prefers open farmland, meadows and coastal fields. Migratory.

Crested Lark *Galerida cristata* 18cm

Similar to Skylark but easily recognised by its prominent, pointed crest. Upperparts are streaked greyish-brown, and underparts are pale buff with streaking on the breast. The bill is sturdy and slightly curved, and the legs are pinkish. In flight, shows broad wings with subtle streaking and a short tail with rusty-tinged outer feathers.

Where to see Rare visitor in southern Finland. Occasionally appears in open areas, fields, grasslands, industrial zones and even urban parks at any time of year.

Shore Lark *Eremophila alpestris* 18cm

About the size of Skylark. Pale brown and streaked above, white below, with a yellow face, black mask and breast-band, and small black feather horns on the head. Females and juveniles are paler with less distinct markings. In flight, it shows dark wings and a long, blackish tail.

Where to see Rare breeder in the fells of northern Lapland. In late autumn and winter, occasionally appears in southern Finland on open fields and wasteland in coastal areas.

Sand Martin *Riparia riparia* 13cm

The smallest of the swallow species. Upperparts are dull grey-brown, underparts white with a clear brown breast-band. The tail is short and only slightly forked. Flight is quick and agile, with light, fluttering wingbeats as it catches insects near water or over open ground.

Where to see A common summer visitor breeding throughout Finland. Nests colonially in sandy banks, gravel pits or riverbanks. Often seen flying low over lakes, rivers and fields, frequently together with other swallows.

Red-rumped Swallow *Cecropis daurica* 17cm

Similar to Barn Swallow, with long tail streamers, but easily recognised by its rusty-orange rump, which becomes paler toward the tail. The face and neck share the same warm tone. Underparts are creamy with faint streaks on the breast, and the vent is black. In flight, it shows pale underwings with dark flight feathers, creating strong contrast.

Where to see: A rare visitor in southern Finland, often found among flocks of Barn Swallows over open areas, lakes or fields.

Barn Swallow *Hirundo rustica* 19cm

Largest of the swallows, with glossy dark blue upperparts, very long tail streamers and a brick-red face and throat contrasting with pale underparts. Tail feathers are dark with white spots. Juveniles show a shorter tail and a paler orange face. In flight, the deeply forked tail and agile, graceful movements are distinctive.

Where to see A common summer visitor throughout Finland, except in northern Lapland. Closely associated with humans, nesting in barns, sheds and bridges, and often seen flying low over fields or water hunting insects.

House Martin *Delichon urbicum* 14cm

Smaller and stockier than Barn Swallow, with glossy dark blue upperparts, a bright white rump, and clean white underparts and face. The tail is short and slightly forked. The white rump is very noticeable in flight, which is smooth but less agile than Barn Swallow's.

Where to see A widespread summer visitor across Finland. Common in towns, villages and rural areas where buildings offer ledges or eaves for nesting. Often seen flying over lakes, rivers and fields catching insects, or collecting mud for nest construction.

Icterine Warbler *Hippolais icterina* 13cm

A medium-sized warbler, larger than Willow Warbler, with a yellowish belly and greyish back. Shows pale wing panels formed by the light edges of the primaries, which stand out against the darker wings. The tail is straight-edged with narrow white outer feathers. Wings are long, with a noticeable primary projection.

Where to see Summer visitor in southern Finland. Breeds in bright deciduous forests, riverside groves, old parks, gardens and other broadleaf-dominated woodlands. Often difficult to spot, but its melodic, fast-paced song helps locate it during the breeding season.

Booted Warbler *Iduna caligata* 12cm

Small warbler, about the size of Willow Warbler, resembling a typical *Phylloscopus* species. Upperparts grey-brown, underparts paler with rusty-buff flanks. Bill short, legs brownish-pink with darker toes. Short primary projection gives a compact appearance. Pale, understated supercilium. The song is a varied, scratchy warble, often including mimicry.

Where to see: Eastern migrant and uncommon breeder in southeast Finland. Prefers bushy and scrubby habitats, often found in farmland thickets, hedgerows and overgrown clearings. Most easily located by its distinctive song during the breeding season.

Sedge Warbler *Acrocephalus schoenobaenus* 12cm

Small, buff-brown warbler with a broad, pale supercilium, dark-streaked crown and heavily streaked back. Rump is plain yellow-brown. Slender bill and brown legs. Wings show a long primary projection, giving a slim appearance; tail short. The song is loud, fast and energetic, a varied chattering series, less melodic than Reed Warbler.

Where to see Summer visitor, breeding widely in Finland. Most common in the south but also reaches Lapland. Prefers marshes, reedbeds, willow thickets and other wet, bushy habitats. Often located by its distinctive song during the breeding season.

Blyth's Reed Warbler *Acrocephalus dumetorum* 13cm

Small, plain brown warbler, very similar to Reed and Marsh Warblers. Upperparts are greyish-brown, underparts pale buff with little contrast. Short primary projection and faint pale supercilium in front of and just behind the eye. Bill long, forehead flat and legs often dark. Best identified by its song: fast, varied and full of mimicry, unlike the repetitive Reed Warbler.

Where to see Locally common breeder in southern and central Finland. Prefers bushy wetlands, reedbeds, meadows and scrubby areas. Often located by its distinctive, melodious singing during the breeding season.

Marsh Warbler *Acrocephalus palustris* 14cm

Small, plain brown warbler, very similar to Reed Warbler, often nearly impossible to separate in the field without hearing its song. Slightly rounder head, shorter bill and greyer-brown back give a softer facial expression. Flanks washed yellowish-brown and rump paler with a warm tone. Primary projection long, with pale tips more obvious. Song is the key to identification: rich, varied mimicry of many other birds, fast and fluent.

Where to see Uncommon breeder in southern Finland. Found in reedbeds, bushy meadows, roadsides, wastelands and overgrown gardens.

Reed Warbler *Acrocephalus scirpaceus* 13cm

Small, warm brown warbler with uniform grey-brown upperparts and paler buff underparts. Lacks strong markings, with only a faint supercilium in front of the eye. Bill is fairly long, legs brownish, and wings long and pointed, giving a slim appearance. Overall, plain and warm-toned. The song is rhythmic and repetitive, a series of chattering phrases, less varied than Marsh or Blyth's Reed Warblers.

Where to see Local breeder in southern Finland. Mainly found in reedbeds alongside lakes, rivers and wetlands, often well-hidden among dense vegetation.

Great Reed Warbler *Acrocephalus arundinaceus* 18cm

Much larger than other European warblers, nearly thrush-sized. Strong, heavy bill, broad head and upright posture when singing. Upperparts warm brown, and underparts whitish with buff flanks and undertail. Short, pale supercilium and long, rounded tail. Usually clambers awkwardly through reeds. Song is unmistakable: loud, harsh and varied, with grating notes and deep phrases, carrying far across wetlands.

Where to see Rare breeder in southern Finland, in large reedbeds alongside lakes and rivers. Most easily heard during spring when males sing prominently.

River Warbler *Locustella fluviatilis* 15cm

Medium-sized, slightly larger than Grasshopper Warbler, with diffuse breast and throat streaking extending onto the flanks. Upperparts olive-brown, and underparts greyish with pale belly and vent. Tail broad and square-ended; supercilium short and indistinct. Usually stays low in dense vegetation, moving quietly and inconspicuously. The song is distinctive: a long, mechanical, cricket-like trill, very even and continuous, often delivered from deep cover.

Where to see Scarce breeder in southern and central Finland. Prefers damp meadows, bushy wetlands, riversides and marshy clearings.

Savi's Warbler *Locustella luscinioides* 14cm

Medium-sized, similar in size to Marsh Warbler. Plain brown with warm reddish-brown upperparts and greyish-buff underparts, faintly mottled on the breast and flanks. Tail broad and rounded with slightly paler tips, often held cocked. Supercilium short and weak, overall inconspicuous. Very secretive, staying low in dense reedbeds and wet vegetation. Song is a continuous, fast, reeling trill, lower and harsher than Grasshopper Warbler, often heard at dusk.

Where to see Summer visitor. Rare breeder in southern Finland, inhabiting dense reedbeds and marshy lakesides.

Grasshopper Warbler *Locustella naevia* 13cm

Small, secretive warbler. Upperparts warm brown with heavy dark streaks; underparts buffish. Tail long and rounded, often flicked when moving through dense cover. Supercilium faint, bill fine and straight. Very elusive, usually glimpsed briefly while creeping low in vegetation. Song is unmistakable: a long, insect-like reeling trill, reminiscent of a grasshopper, often delivered at dusk or during the night.

Where to see Local breeder from southern Finland to the Oulu region, mainly in marshes, wet meadows and scrubby habitats.

Wood Warbler *Phylloscopus sibilatrix* 12cm

Small, similar in size to Willow Warbler. Upperparts are bright green, contrasting with white underparts and a yellow throat and breast. Bright yellow supercilium on the head, lemon-yellow throat and white belly. Bill fine, legs pale and primary projection longer than in Willow Warbler or Chiffchaff. Tail is relatively short. The song is distinctive: a clear series of descending whistles followed by a fast, insect-like trill.

Where to see Summer visitor. Fairly common breeder from southern Finland to the Oulu region, uncommon in Lapland. Prefers mature deciduous and mixed forests.

Yellow-browed Warbler *Phylloscopus inornatus* 10cm

Small, active warbler, smaller than Willow Warbler. Upperparts moss-green, underparts pale. Bright, well-defined yellow supercilium and fine pointed bill. Two broad wing-bars, reminiscent of Pallas's Leaf Warbler. Tail is often flicked while bird moves through low foliage. Very restless, usually seen in mixed flocks of small birds. Vocal, giving a high-pitched, thin call *tsoeest*.

Where to see Rare autumn visitor, mainly along coasts and on archipelagos. Prefers deciduous forests, small groves and low shrubs, often near water.

Pallas's Leaf Warbler *Phylloscopus proregulus* 9cm

Small, agile warbler, slightly larger than Goldcrest. Upperparts pale green, with a long supercilium and a prominent central crown stripe. Head relatively large, tail short. Pale yellow rump visible in flight, and wing-bars noticeable. Moves actively, often in mixed flocks with Goldcrests and other warblers. The call is a sharp *tsui*.

Where to see Rare autumn visitor, mainly along coasts and islands. Prefers reedbeds, marsh edges, low shrubs, parks and gardens, usually in dense vegetation where it can move quickly and remain inconspicuous.

Dusky Warbler *Phylloscopus fuscatus* 14cm

Larger than Chiffchaff, compact with rounded wings. Upperparts dark grey-brown, underparts paler brown, with no yellow on flanks or undertail. Long pale supercilium contrasts with otherwise plain, dusky plumage. Bill thin and sharp; legs slightly long, giving a slender appearance. Similar to Chiffchaff but darker, more uniform, with shorter primary projection. Call is a soft, sharp *teck*, reminiscent of Lesser Whitethroat.

Where to see Rare eastern visitor, mainly in autumn. Occurs in coastal scrub, bushes and thickets, often joining mixed flocks of small birds.

Willow Warbler *Phylloscopus trochilus* 12cm

Small, slightly larger than Goldcrest. Upperparts greenish-brown, underparts pale, sometimes yellowish. Supercilium faint but visible, bill fine and pointed, and legs usually pale (sometimes darker, like Chiffchaff). Often tricky to separate from Chiffchaff without hearing its song, but shows longer primary projection, roughly the same length as the tertials. The song is a sweet, descending, melodic warble, often trilled and repeated, especially in spring and summer. The call is a soft whistle, *huitt*.

Where to see Common breeder throughout Finland. Found in deciduous and mixed forests, parks and gardens, often near open areas with shrubs or low trees.

Chiffchaff *Phylloscopus collybita* 11cm

Upperparts are olive-brown, underparts pale with faint brownish tones. Faint supercilium, fine pointed bill, and often darker legs. Hard to separate from Willow Warbler if silent but shows shorter primary projection. Very active, usually moving through low foliage or perching openly. The song is a simple, repetitive *chiff-chaff* phrase, repeated frequently.

Where to see Common breeder throughout Finland. Found in deciduous and mixed forests, parks and gardens, often near shrubs or low trees.

Greenish Warbler *Phylloscopus trochiloides* 10cm

Slightly smaller than Willow Warbler. Upperparts olive-green, underparts pale. Pale, distinct supercilium reaches the nostril. Bill is fine and pointed, legs greyish-brown. Shows a short, thin pale wing-bar and short primary projection. The song is a high, thin, fast warble, slightly buzzing and ascending, often repeated.

Where to see Breeds locally in southern and central Finland, uncommon north of the Kuusamo region. Prefers deciduous forests, small groves and low shrubs, often near water or along open woodland edges.

Arctic Warbler *Phylloscopus borealis* 12cm

Upperparts olive-green, underparts pale with greyish breast and whitish throat. Supercilium is pale but subtle, not extending to the nostril. Bill is fine and pointed. Short, thin, pale wing-bar present; primary projection longer than Greenish Warbler. The song is a fast, high-pitched trill, slightly buzzing and frequently repeated.

Where to see Rare summer visitor in Lapland, with a few breeding pairs. Found in birch and mixed forests, groves and low shrubs, often near water.

Garden Warbler *Sylvia borin* 14cm

Medium-sized and rather featureless warbler. Upperparts brownish-grey, underparts paler greyish-white. Lacks obvious markings. Absence of eye-ring, supercilium or crown patch give a very anonymous appearance compared to other warblers. Bill slim, legs greyish. Can be confused with female Blackcap but lacks the brown cap. The song is a rich, fast, melodious warble, less fluty and more continuous than Blackcap.

Where to see Fairly common summer breeder in southern and central Finland. Favours woodland edges, clearings, gardens and shrubby habitats.

Barred Warbler *Curruca nisoria* 16cm

Large warbler, noticeably bulkier than Common or Lesser Whitethroats. The adult has a grey head, back and tail; underparts whitish with strong grey bars on breast and flanks. Juveniles are browner with faint streaks instead of bars. Adults have yellow eyes, while juveniles' eyes are dark. The song is harsh and scratchy, mixed with whistles, sounding like a rougher version of Common Whitethroat.

Where to see Rare and local summer breeder in southeastern Finland. Prefers dense scrub, thickets and overgrown field margins.

Common Whitethroat *Curruca communis* 14cm

Medium-sized warbler with a grey head, white throat and reddish-brown wings contrasting with a brown back. Underparts are pale, often buff-tinged on the flanks. The tail is long, with white outer feathers and dark central ones, frequently flicked. Females and juveniles are duller, with browner heads and less contrast. Song is a fast, scratchy warble, typically delivered from a bush or during short display flights.

Where to see Summer visitor. Fairly common breeder throughout Finland, favouring bushy areas, meadows, field edges and clearings.

Lesser Whitethroat *Curruca curruca* 12cm

Smaller and plainer than Common Whitethroat. Grey head and nape, brown back and white underparts. Throat is clean white, contrasting with grey face; wings and tail are plain brown. The tail is often flicked but lacks the white outer feathers seen in Common. Both sexes appear similar; juveniles are slightly duller. Song is a short, rattling trill, usually heard before the bird is seen, unlike the scratchy warble of Common.

Where to see Summer visitor. Fairly common breeder in southern and central Finland, favouring scrub, hedges and woodland edges.

Blackcap *Sylvia atricapilla* 14cm

Medium-sized warbler, overall plain grey with a subtle olive tinge on wings and back. Male is unmistakable with a neat black cap, while female and juvenile show a chestnut-brown cap. Underparts are paler grey, throat whitish. Bill is slim, legs greyish. The song is rich, varied, and flute-like, considered one of the most beautiful among warblers; call is a sharp *tac-tac*. Very active and often found in dense foliage.

Where to see Fairly common summer breeder in southern and central Finland. Prefers mixed woodland, parks, gardens and bushy habitats. Some individuals visit bird feeders during winter in the south.

Bohemian Waxwing *Bombycilla garrulus* 20cm

Size of a Starling, with a prominent crest and silky, colourful plumage giving a majestic look. Overall pale brown with a rusty-brown face, thick neck and strong black mask through the eyes. A large black bib under the bill. The tail has a broad yellow band, the primaries have striking V-shaped tips and the bright red wax-like tips on the wings are unmistakable. Call a high-pitched, thin, trilling *sreee*, often heard in flight or from treetops.

Where to see Migratory bird, often seen in large flocks in autumn and winter in berry-rich areas, gardens and city centres. Breeds in northern coniferous forests.

Firecrest *Regulus ignicapilla* 9cm

Similar in size and shape to Goldcrest. It has a bold white eyebrow, a black line through the eye and a small white spot beneath the eye, giving a strong contrast to the yellowish-green back. A distinct white wing-bar is also prominent. Male shows an orange crown stripe bordered by black, while the female's is yellow, as in Goldcrest.

Where to see Uncommon visitor, with records increasing each year and some breeding noted. Most often seen from the Oulu region southwards, in habitats similar to Goldcrest.

Goldcrest *Regulus regulus* 9cm

Tiny bird with olive-green plumage. Male has a bright orange crown bordered by black; the female's crown is yellow. Both sexes feature a white eye-ring, distinct white wing-bar and pale grey belly. Spring song is very high-pitched and thin, a rapid series of notes, often difficult to hear.

Where to see Common breeder from southern Finland north to southern Lapland, less frequent further north. Prefers coniferous forests, especially spruce, often staying high in the treetops and moving actively through the canopy.

Nuthatch *Sitta europaea* 13cm

Compact, neckless bird with a large head and long, pointed bill. Upperparts are light blue-grey, underparts whitish with a reddish-brown undertail marked with white spots. A long black line runs through the eye to the back of the head. Birds of the race *asiatica* show more white above the eye; *europaea* females have a rusty-buff vent. Often climbs head-down on tree trunks. Call is a sharp, ringing *twit* or *twit-twit,* repeated rapidly.

Where to see Uncommon breeder in southern Finland; *europaea* breeds locally; *asiatica* occurs as a migrant from Siberia. Found in deciduous forests, parks and at feeders in winter.

Treecreeper *Certhia familiaris* 13cm

Small bird with a brown-streaked back, white underside, long down-curved bill, long tail and prominent white supercilium. Typically creeps up tree trunks before flying to the next tree. Call is a repeated sharp *srri*; song is a high-pitched, thin, slightly warbling series of notes, usually delivered from a high branch.

Where to see Uncommon breeder from southern Finland to the Oulu region, rarer further north. Nests between tree trunks and loose bark. Found in old coniferous, deciduous and mixed forests. Mainly resident; partial migrant in northern and eastern areas.

Wren *Troglodytes troglodytes* 9cm

Tiny, plump bird with a short neck and upright tail, making it appear very small. Upperside reddish-brown, underside dirty brown with faint streaks. Shows a pale eyebrow and relatively long bill. Alarm call is a sharp *zerrr*; song is a loud, high-pitched series of trills and rattles: *zi-zi-zi-zirrrr, svi-svi-svi-zerrr, sivi*.

Where to see Uncommon breeder, more frequent from southern Finland to the Oulu region, scarcer further north. Prefers spruce-dominated coniferous and mixed forests, stream banks, forest glades and edges of logging areas with woodpiles and dense thickets.

Dipper *Cinclus cinclus* 19cm

Compact, Starling-sized bird with short tail. Dark brown head and belly contrast with black back and large white bib, making it unmistakable. Juveniles are grey with stripes. In flight, wingbeats are rapid, especially over rivers. Call is a sharp, electric *zrik*; song is a soft, high-pitched warble, often sung from rocks or low branches near flowing water.

Where to see Winter visitor and uncommon breeder, mostly in northern Finland. Breeds along clean, fast-flowing streams and rivers with rocky banks or man-made nesting sites. In winter, it may occur nationwide along rivers and streams.

Common Starling *Sturnus vulgaris* 21cm

juv.

Glossy dark plumage shows green and purple iridescence. Feathers are often spotted, especially on the back. Bill yellow with a bluish-grey base; juveniles light brown with paler throat and darker bill. Call is a short, sharp *churrr*.

Where to see Summer visitor. Breeds in parks, gardens and farmland, common from the south to the Oulu region, scarcer further north. In autumn, large flocks form of hundreds or thousands of birds.

Rose-coloured Starling *Pastor roseus* 21cm

Structurally similar to Common Starling, but adults have black wings and tail, pinkish-rose back, upperparts, and belly, and a black head with a small crest. Bill and legs are pale pink. Juveniles are pale brown with darker wings and a pale bill base, with paler upperparts noticeable in flight.

Where to see Uncommon visitor. Occasionally observed in flocks with Common Starlings during spring or autumn migration. Adults may sometimes be seen in summer, especially in southern Finland.

Mistle Thrush *Turdus viscivorus* 28cm

Large thrush with grey-brown back, pale cheek patch, and rounded dark spots on the white chest and belly. Perches upright, appearing more stately than smaller thrushes. In flight, shows white underwings and a long tail with white corners, distinguishing it from Song Thrush.

Where to see Locally common breeder throughout Finland, except northernmost Lapland. Prefers open pine forests, pine-dominated mixed forests and sometimes open patches within spruce forests. During migration, often seen in fields with flocks of other thrush species.

Song Thrush *Turdus philomelos* 21cm

Slightly smaller than Mistle Thrush. Brown back and face, yellow-buff breast with distinct arrowhead-shaped brown spots extending to the belly. In flight, short tail and rusty-buff underwings are visible. Song is clear, fluting and melodic, a series of short, high-pitched phrases often ending on a rising note.

Where to see A common breeder throughout Finland, mainly in coniferous forests. During migration, often visits yards, parks and fields while feeding. Migrates at night in autumn, usually alongside Redwings.

Redwing *Turdus iliacus* 21cm

Similar in size to Song Thrush. Brown back, white belly and breast with dark streaks, and rusty-red flanks. Distinct white supercilium and submoustachial stripe. In flight, short tail and striking rusty-red underwings are visible. Song is a high-pitched, thin, melodic series of short phrases, often rising at the end. Call is a sharp *seep* or *tseep,* especially during nocturnal migration.

Where to see Common breeder at forest edges in young coniferous areas, willow thickets and mountain birch forests. Feeds in fields with other thrushes during spring migration.

Fieldfare *Turdus pilaris* 25cm

Grey head and lower back, with brown back and wings between the greyish areas. White supercilium, ochre-coloured breast with black streaks and an orange bill. In flight, it shows a longish black tail, white underwings and brown upperwings. Call is often chattering a *schack-schack-schack*.

Where to see Common breeder throughout the country. Prefers field edges, parks, yards and gardens. In short, any area with nearby open spaces such as fields or lawns. During spring migration, often seen feeding in fields alongside other thrushes.

Blackbird *Turdus merula* 26cm

Male is entirely black with a bright yellow bill and eye-ring; female and juvenile are brown with spotted breast and paler throat. Often hops on the ground in search of worms. The song is melodious, loud and fluting, with a wide variety of calls; during migration, a sharp *srrri* is common.

♂

Where to see Common breeder from the south to Oulu region, scarcer further north. Prefers mixed and deciduous forests with some spruce and also thrives near humans in gardens and parks.

♀

Ring Ouzel *Turdus torquatus* 26cm

Similar in size and shape to Blackbird. Male is mostly black with a white crescent on the chest and pale wing patches, which are visible on the ground and in flight. The female is brownish with a paler crescent and similar pale wings. The song consists of a series of clear, fluty phrases: *tuu-tuu-tii-tii*.

♂

Where to see Uncommon breeder in northern Lapland. Migrates through Finland and can be observed across the country during spring migration, often on open hillsides and upland areas.

♀

Spotted Flycatcher *Muscicapa striata* 14cm

A small, slender bird with grey-brown upperparts, white underparts and fine streaks on the forehead, cheeks and breast. Sexes are alike, both appearing plain and delicate. Often perches upright on exposed branches or posts, darting out to catch insects in quick, agile flights before returning to the same spot. Not very vocal; typical call a sharp, dry *zee*.

Where to see A common breeder across Finland. Found in gardens, parks, open forests, clearings, forest edges and along field or shore margins.

Robin *Erithacus rubecula* 13cm

Small, compact bird with a plump body and large, dark eyes. Adults show brownish upperparts, a striking orange-red face and breast bordered with grey, and a whitish belly. Juveniles are mottled brown with no red, appearing speckled overall. Moves with short, quick hops, often pausing upright on low branches. Call a sharp, repeated *tick*. Song is a soft, melodious warble with fluid, varied phrases.

Where to see Common breeder across Finland except northern Lapland. Prefers spruce, deciduous and mixed forests with dense undergrowth; also frequent in gardens, parks and forest edges.

Thrush Nightingale *Luscinia luscinia* 16cm

Medium-sized songbird with warm reddish-brown upperparts and greyish underparts. Breast diffusely mottled grey-brown, upperparts less reddish than Common Nightingale. Tail long and often slightly raised; bill slender and dark. Sexes similar, but juveniles duller and more mottled. Song is brilliant, rich and varied, with whistles, trills and deep notes, often performed at dusk or night.

Where to see Summer visitor and fairly common breeder in southern and central Finland. Prefers dense thickets, riversides, reedbeds and overgrown clearings. Shy and best located by song.

Bluethroat *Luscinia svecica* 13cm

♂

A small, Robin-sized bird. The adult male is unmistakable, with a vivid blue bib featuring an orange spot, and bold white eyebrows. Beneath the bib are contrasting black and orange markings. All plumages show a black tail with a distinctive rusty-red base, often flicked or held upright. Females lack the strong blue but display clear eyebrows and moustachial stripes, and a dark breast-band. The song is loud, varied and melodious, often incorporating mimicry.

♀

Where to see Uncommon breeder in Lapland, favouring moist bushes and forests, especially willow and mountain birch. Territories are often near bogs or water bodies. During migration, found across Finland, including archipelago islands.

Red-breasted Flycatcher *Ficedula parva* 11cm

Smaller than Pied Flycatcher. An adult male has a rounded grey head, large dark eyes and a noticeable white eye-ring, bright orange throat and upper breast, brown back and whitish belly. Tail dark with white-edged outer feathers. Females and juveniles lack the orange, showing a pale buff throat and breast, brown upperparts and whitish underparts. Song is brilliant, a rapid series of thin, high-pitched notes with delicate trills; call a sharp *tek*.

Where to see Scarce breeder in southern Finland. Prefers light, mixed forests with mature deciduous trees. More widespread on migration.

Pied Flycatcher *Ficedula hypoleuca* 13cm

Male black above, white below, with a white wing patch, a small primary patch and a white spot above the bill; amount of white varies between individuals. Female warm brown above, whitish below, with smaller but distinct wing patches. Active and vocal, the song lively and varied, with short whistles and trills.

Where to see Common breeder across Finland. Frequently nests in nest boxes in gardens, courtyards and parks, as well as in natural tree cavities.

Collared Flycatcher *Ficedula albicollis* 13cm

Similar to Pied Flycatcher but shows more white overall. Most distinctive features are a broad white neck-ring, extensive white on the rump, a larger white patch above the bill, and a bigger wing and primary patch. Individual variation is considerable. Females resemble female Pied but are often paler brown with a larger primary patch.

Where to see Uncommon visitor. Most records are from archipelago islands during spring migration, though occasionally observed inland, especially in southern Finland.

Red-flanked Bluetail *Tarsiger cyanurus* 13cm

♂

About the size of Pied Flycatcher. The adult male is unmistakable, with brilliant blue upperparts, a pure white throat and belly, and vivid orange flanks, creating a brilliant contrast. Female and juvenile males are brownish overall but retain orange flanks and a bluish tail, showing a subtler version of the male's pattern. Song is a simple, melodious phrase, while calls include a sharp *hvuit-hvuit* or dry *tack-tack-tack*.

Where to see Summer visitor. Breeds mainly in eastern Finland, especially around Kuusamo, in coniferous forests, edges and lush undergrowth.

♂

juv.

Black Redstart *Phoenicurus ochruros* 13cm

♂

♀

Similar to Common Redstart, with a slim body, upright posture and frequent tail-flicking. The adult male has grey-black plumage, black face and breast, bright orange-red tail and a small white wing patch. Females and juveniles are plainer, ash-grey with a reddish tail. Song begins with short, clear whistles, followed by scratchy, buzzing phrases, often ending with a series of trills, sometimes rendered as *si-srue-til-ill-ill-ill*. Calls include a sharp *tchik* or soft *tsit*.

Where to see Uncommon breeder. Nests mainly in industrial areas, railway yards and urban sites. During migration, it is fairly common in the archipelago.

Common Redstart *Phoenicurus phoenicurus* 13cm

Small, slim, Robin-sized bird with colourful plumage. Adult males have a slate-grey back, black face and throat, bright white forehead and vivid orange-red breast, belly and tail. Tail is constantly flicking, making the bird very noticeable. Females and juveniles are duller, with brownish upperparts and warm orange tail, but retain the characteristic tail-flicking. Song is a clear, melodious high-pitched warble; call a sharp *huit* or soft *tchik*, often given during flight or when alarmed.

Where to see Summer visitor. Prefers open forests, especially pinewoods, but also heathland forests, parks and gardens. Nests in tree-holes or nest boxes.

Whinchat *Saxicola rubetra* 13cm

Small, upright bird with a short tail and slender bill. The breeding male is brilliant, with a warm brown back streaked with darker markings, pale orange-buff breast, white belly and a bold white supercilium contrasting with the dark eye-line. Female and juvenile are duller, with a less distinct facial pattern. The tail is black with a white base and outer edges. Often perches on shrubs, fence posts or tall vegetation, flicking tail and wings. Call is a sharp, thin *tsip*, while the song is a pleasant, melodious warble, delivered from an elevated perch.

Where to see Summer visitor and locally common breeder across Finland, except far northern Lapland. Prefers open meadows, pastures and low scrub.

Siberian Stonechat *Saxicola maurus* 12cm

Very similar to Common Stonechat and sometimes tricky to identify but paler overall. Breeding male shows a small orange breast-patch, large white patches on the nape and wings, and a prominent white rump above the black tail. Underwing has black elbow feathers. Female and juveniles are duller, with buff-orange breast, streaked underparts and a plain rump. Often perches on shrubs or posts, flicking tail and wings. Call is a sharp, thin *chack*, like Common Stonechat.

♂

Where to see Very rare vagrant in Finland, mostly occur south of Oulu and the Kuusamo region during migration.

Common Stonechat *Saxicola rubicola* 12cm

♂

♀

Similar in body shape to Whinchat. The breeding male has a black head and throat, bright orange breast, white collar patch, brown back with darker streaks, and white underparts. Female and juvenile are duller, with buff-orange breast and brown head and back with streaks. Often perches conspicuously on fence posts, shrubs or rocks, flicking wings and tail. Call is a sharp, whistling *chack*, and song is a short series of melodious phrases delivered from exposed perches.

Where to see Uncommon visitor, favouring open fields, pastures, coastal meadows and low scrub.

Northern Wheatear *Oenanthe oenanthe* 15cm

Male has a grey head and back, light brown throat, white supercilium and black mask. Underparts are white, legs black. Female is beige-brown overall, darker on the back, with a faint black line between bill and eye and a white supercilium. Bill is thin, short and pointed. In flight, shows black-and-white tail with a characteristic black T-pattern and white edges. Moves nimbly on the ground in an upright posture. Voice is a soft, melodic *chit* or *tseep*, often given in flight or from a low perch.

Where to see Locally common summer visitor. Breeds throughout Finland, from archipelago islands to Lapland mountains, favouring dry open areas, clearings, rocky shores and stone piles.

Dunnock *Prunella modularis* 14cm

Size of a Robin. Plumage is mainly brown and streaked, with a greyish head, brown ear-coverts and brown crown. Bill is thin and warbler-like, giving a delicate, inconspicuous appearance. Legs are slender and dark. Often moves quietly on the ground or among low shrubs, but may sing from treetops, especially in spruce forests. The song is a rapid, continuous series of high-pitched notes and trills, sometimes broken by short pauses, creating a soft, tinkling melody.

Where to see Summer visitor, breeding mainly in southern Finland but present nationwide. Prefers lush spruce forests, mixed forests and forest edges with low conifers for nesting.

Tree Sparrow *Passer montanus* 13cm

Slightly smaller than House Sparrow, with slimmer build and more active behaviour. Distinctive chestnut-brown crown and nape, contrasting with white cheeks marked by a neat black spot. Black bib is smaller than in House Sparrow. Back brown streaked darker, wings with white bars, underparts pale grey-brown. Sexes are alike. Juvenile duller but already shows cheek-spot. Call a sharp, cheerful *chip-chip*.

Where to see Widespread but local in Finland, favouring farmland, orchards, villages and gardens.

House Sparrow *Passer domesticus* 15cm

♂

♀

Sturdy, slightly larger than Tree Sparrow, with thicker bill and more robust build. Male distinctive with grey crown, black bib and throat, chestnut nape, pale cheeks, brown streaked back and grey underparts. Female and juvenile duller, sandy-brown with streaked back, grey-brown underparts and pale eye-stripe, lacking bib and crown pattern. Call a harsh *chup*, often in chattering groups.

Where to see Resident. Common in southern and central Finland. Strongly tied to human settlements, farms, towns and city centres.

Grey Wagtail *Motacilla cinerea* 18cm

Slender, sparrow-sized wagtail with long tail, constantly wagged. Adult male in breeding plumage has grey back, bright yellow underparts and black throat patch; head mostly grey with white supercilium. Female and juvenile paler yellow below, less contrasting grey above, without black throat. Often seen running along water edge, feeding on insects. Call a soft, high-pitched *tsi-weet* or thin *tsip*.

Where to see Uncommon breeder. Breeds along rivers, streams, ponds and wetlands across Finland; can also be seen during migration.

Yellow Wagtail *Motacilla flava* 16cm

Similar in shape to White Wagtail, though slightly smaller. Breeding male has bright yellow underparts, a greenish back and a greyish head; some subspecies show more olive or grey tones. The *flava* subspecies displays a distinct white supercilium. Female and juvenile are duller, with paler yellow underparts and less distinct head markings. Call a soft, high-pitched *tsip-tsip*; song a pleasant, twittering warble, often given in flight or from low perch.

Where to see Summer visitor. Fairly common breeder across Finland, especially in wet meadows, riversides, farmland and open fields. Southern populations are mainly *flava*, northern mainly *thunbergi*.

Citrine Wagtail *Motacilla citreola* 16cm

♂

Similar in size and shape to Yellow Wagtail, with a long tail constantly wagged. Adult breeding males are striking, with bright lemon-yellow head, throat, breast and belly contrasting with grey back, blackish wings with pale edges, and a long dark tail with white outer feathers. Females and non-breeding birds are duller, with a whitish face and paler yellow underparts. Juveniles are grey-brown with faint yellow tones. Call a sharp, high *tsree*.

♀

Where to see Very scarce breeder in southern Finland, favouring shallow wetlands, marshes and wet meadows; also a rare migrant elsewhere.

White Wagtail *Motacilla alba* 18cm

♂

Distinctive for its long tail, constantly wagged up and down. Plumage black, grey and white: black throat patch and crown, white face, grey back and white belly. Juveniles are similar but paler, with less contrasting black on head and throat. Slim, agile and long-legged, often seen running on ground or perching on stones and fences. Call a sharp, high-pitched *chissick*.

♀

Where to see Very common breeder across Finland, favouring open areas, riverbanks, fields, meadows and urban habitats.

Richard's Pipit *Anthus richardi* 19cm

Large, upright pipit with long legs and tail, between the size of a Skylark and a Starling. Upperparts are sandy-brown with darker streaks, underparts pale buff or whitish with fine streaking on the breast. Head plain with a bold, pale supercilium; bill long and pointed. Tail has white outer feathers, often flicked. Call a sharp, sparrow-like *shreep*, usually given in flight.

Where to see Rare autumn migrant in Finland, mainly along coasts. Favours open fields, meadows and grassy shorelines.

Tree Pipit *Anthus trivialis* 15cm

Medium-sized pipit with streaked brown upperparts and pale underparts. Breast and flanks show fine streaking; belly whitish. Head with clear pale supercilium and slender pointed bill. Very similar to Meadow Pipit, but slightly heavier, with bolder breast streaking, more contrasting face pattern and shorter, more curved hindclaw. Perches on trees or bushes more often than Meadow. Song distinctive rising series of notes delivered in display flight, ending with trills.

Where to see Common summer breeder across Finland, in woodland edges, clearings and forest meadows.

Meadow Pipit *Anthus pratensis* 14cm

Small, slim pipit with brown-streaked upperparts and pale underparts. Breast and flanks finely streaked; belly whitish. Head rather plain, with thin pale supercilium and slender bill. Very similar to Tree Pipit but smaller and slimmer, with less contrasting face and finer breast streaking. Tail longish with white outer feathers, often pumped. Hindclaw longer and straighter. Song a thin, repeated *seep-seep*, delivered in fluttering song flight.

Where to see Common breeder across Finland, especially in open grasslands, fells, marshes and meadows.

Rock Pipit *Anthus petrosus* 16cm

Medium-sized, dark and rather bulky pipit. Plumage grey-brown above, underparts dirty-buff with diffuse streaking on breast and flanks. Head rather plain, supercilium weak, and bill dark and fairly strong. Tail dark with white outer feathers less obvious than in other pipits. Flight low and direct, often along rocky shorelines. Call a sharp, slightly rough *vist*.

Where to see Uncommon breeder along rocky coasts, particularly in southern and western coastal areas. Also seen outside the breeding season along shores and tidal flats.

Red-throated Pipit *Anthus cervinus* 14cm

Slender pipit, with longish tail and thin bill. Breeding adult has a reddish throat, face and breast, contrasting with heavily streaked brown upperparts and whitish belly. In non-breeding adult and juvenile plumages, throat is pale buff and streaking still strong. Shows white outer tail-feathers in flight, like other pipits. Call a sharp, buzzing *dzeet,* distinctive when migrating flocks pass overhead.

Where to see Breeds locally on tundra and fells in northern Lapland. During migration, seen more widely across Finland, often in open fields and marshy meadows.

Brambling *Fringilla montifringilla* 15cm

♂

♀

Similar to Chaffinch. All plumages show rusty-orange throat, breast, shoulder feathers and wing-bars, with a white rump, blackish tail and black-spotted flanks. Adult males in summer have a shiny black head, neck and back, with a black bill. Females, juveniles and winter males have a greyer head with a yellowish bill. Song and call include a dry Greenfinch-like *dzaa* and a rough *rrrryh*.

Where to see Common breeder in Lapland, also in the south. During migration, large flocks pass through Finland, often with Chaffinches.

Chaffinch *Fringilla coelebs* 15cm

A medium-sized, thick-billed finch with a slightly longer tail than a sparrow. Males in summer are brightly coloured, with reddish underparts, grey head and neck, brown upper back and greenish lower back. Females and winter males are duller brownish-grey. In flight, two bold white wing-bars are visible, and white outer tail feathers contrast with the black tail. The song is a cheerful, accelerating series of notes, ending in a flourish.

Where to see Very common breeder across Finland in forests, parks and gardens; during migration often forms large flocks with Bramblings.

Hawfinch *Coccothraustes coccothraustes* 17cm

♂

♀

Largest of the finches, bulky with a massive head and thick, conical bill for cracking hard seeds. Size between Starling and Song Thrush. Plumage warm brown, wings darker with bold white wing-bar and tail black and white. The head shows a black bib and eye-stripe, and a pale nape; bill bluish in summer, horn-coloured in winter. Flight strong, with deep wingbeats. Call a sharp metallic *pix*.

Where to see Scarce breeder in southern Finland. Prefers mature deciduous woods, orchards, parks and large gardens. Often shy and secretive.

Common Rosefinch *Carpodacus erythrinus* 14cm

♂

♀

Size of Chaffinch with short, stout bill. Adult male in breeding plumage bright crimson-red on head, breast and rump, contrasting with brown back and wings; belly whitish. Female and juvenile much duller, grey-brown with faint streaking and pale underparts, lacking red entirely. In flight appears plain, with no wing-bars. Call a soft *dyip* or *weet*. Song a simple, melodious warble, often compared to the phrase *nice to meet you*.

Where to see Widespread summer visitor in Finland, breeding in thickets, riversides and parks.

Pine Grosbeak *Pinicola enucleator* 21cm

♂

Large, heavy finch, noticeably bigger than Bullfinch. Body thick and rounded, with stout, curved bill. Male rose-red with grey wings, white wing-bars and black tail. Female and immature yellowish-orange to olive, with grey body and wings. Flight slow and heavy, recalling a thrush. Often tame and sluggish, feeding on buds, berries and seeds. Call a mellow, flute-like *tyu* or soft whistles. Song a clear, warbling series of sweet notes.

Where to see Local breeder in northern Finland, especially in coniferous forests. In winter may move south, often at feeders or eating Rowan berries.

Bullfinch *Pyrrhula pyrrhula* 16cm

♂

juv.

Sturdy, compact finch with short neck and rounded appearance. Bill thick and black, adapted for crushing seeds and buds. Male unmistakable with bright rose-red underparts, black cap, face and tail, grey back and white rump. Female and juvenile similar, but underparts buff-brown instead of red. Wings short and rounded. Call a soft, melancholy *peu*, often repeated. Song quiet and subdued, a mix of whistles and short notes.

Where to see Widespread but uncommon breeder across Finland, in woodlands, parks and gardens. Winters in the south and often visits feeders.

♂ ♀

Greenfinch *Chloris chloris* 15cm

♂

Size of Chaffinch, but stockier with a large head and strong bill. Overall green in appearance. Male brighter, with yellow patches on wings and tail, and greyish cheeks. Female and juvenile duller, more brownish-olive with less yellow. In flight shows broad yellow wing-bars and yellow tail-sides, often giving a heavy, fluttering impression. Call a nasal *dzwee* or *djuu*, often in flight. Song a mixture of twittering phrases and wheezy notes, sometimes delivered in display flight.

♀

Where to see Common breeder across southern and central Finland. Prefers gardens, parks, farmland edges and open woodland.

Twite *Linaria flavirostris* 13cm

This finch has a slim build and a long, notched tail. Bill grey, rather long and pointed. The plain plumage features a grey-brown head, brown back, dark wings with pale wing-bars, and warm buff rump and bib. In summer male shows faint reddish breast wash but lacks bright colours of Redpoll or Linnet. Juvenile similar but duller. Flight undulating, often in small restless flocks. Call a nasal *tveet* or buzzing *twai*.

aut.

Where to see Fairly common in southern Finland, scarcer further north. Prefers farmland, coasts and cultural landscapes.

Linnet *Linaria cannabina* 13cm

♂

♀

Slightly larger than Redpoll, slimmer with longer tail. Bill is grey and fairly slender. Male in summer shows red forehead and breast, brown back with darker streaks, and grey head-sides; wings dark with pale wing-bars and white-edged tail. Female and winter male lack red, and are browner overall with a streaked breast. Juvenile similar to female but plainer. Song a pleasant, musical warble; call a nasal *twit* or rolling *lin-lin*.

Where to see Common breeder in southern Finland, favouring farmland, heaths, coasts and open scrub.

Redpoll *Acanthis flammea* 13cm

arctic

♂

♀

A small finch. Plumage brown and heavily streaked on the head, back and flanks, with a small red forehead patch in all plumages. Adult males in spring display a reddish breast. Bill is short and brownish, with pale feathering at the base. The Arctic form *exilipes* is paler overall, with minimal streaking, while the subspecies *flammea* is smaller, darker and more heavily streaked. Moves actively in flocks, often mixed with other finches. Very vocal, producing a sharp, repeated *chett-chett-chett*.

Where to see Widespread breeder and migrant across Finland, inhabiting parks, gardens, and coniferous and mixed forests.

Two-barred Crossbill *Loxia leucoptera* 15cm

Slightly smaller than Parrot and Common Crossbills. Adult males show bright red plumage with broad white wing-bars, while females are greenish-grey with a boldly spotted back, bright yellowish-green rump and the same broad wing-bars. Both sexes appear slimmer than other crossbills, with a smaller, finer bill. Call is weaker and higher-pitched than Common Crossbill, consisting of a repeated *chip-chip-chip* and a trumpet-type *tviiiht*.

Where to see Uncommon breeder in eastern Finland and Lapland. Numbers fluctuate greatly depending on cone crops; in poor food years, it may migrate south, occasionally appearing almost anywhere in Finland.

Parrot Crossbill *Loxia pytyopsittacus* 17cm

♂

♀

Common/Parrot

A little larger than Common Crossbill, with a bull-necked appearance and stronger head. Adult males are red on the head, underparts and rump, with blackish wings. Females are greenish overall, with a greyish head, blackish wings and yellowish-green rump. The bill is noticeably larger, deeper and more powerful than Common Crossbill, with a strongly curved lower mandible. Call is a hard, repeated *gip-gip-gip*.

Where to see Fairly common throughout Finland. Prefers mature coniferous forests rich in cones, especially pine. Often moves in noisy flocks, seen both in flight and perched high in tree crowns.

Common Crossbill *Loxia curvirostra* 16cm

♂

♂ ♀

All crossbills are similar. Male has a red head, underparts and rump, with blackish wings. Female greenish overall with greyish head, blackish wings and yellowish-green rump. Slightly smaller than Parrot Crossbill, with smaller head and neck, shorter bill and straighter lower mandible. Call also similar to Parrot, given frequently in flight: a loud, metallic *glipp-glipp-glipp*.

Where to see Common breeder throughout Finland. Prefers coniferous forests where cones are abundant. Often moves in flocks and may be seen in flight or perched high in spruce treetops.

Goldfinch *Carduelis carduelis* 13cm

juv.

Adults are unmistakable. Smaller than Greenfinch, with thick pale bill. Plumage very colourful: face dark red, cheeks white, and crown and nape black. Back pale brown and rump white. Wings black with bold yellow wing-bars; tail short and black. Underparts mostly white, with some buff on flanks. Juvenile similar but without red face; head greyish, wings and tail as adult. Call a high, tinkling *ticl-ticl-ticl*, often in flight. Song lively, twittering, with bright trills and notes.

Where to see Common in the south. Prefers parks, gardens and city centres, often in flocks outside breeding season.

Serin *Serinus serinus* 12cm

♂

♀

Small, Siskin-sized finch with a short, dark bill. Adult males show bright yellow on the breast, rump and eyebrows, extending behind the ear toward the neck. Back and flanks are streaked black; belly whitish. Females are similar but paler, with less distinct yellow. Both sexes are finely streaked overall. In flight, wing-bars are smaller than those of Siskin, and tail is entirely dark, lacking yellow outer tail-coverts. Call a sharp, trilling *tzeee*; song a lively, continuous twittering warble, often given in display flight.

Where to see Rare visitor in southern Finland, mainly during spring and autumn migration in parks and gardens.

Siskin *Spinus spinus* 12cm

♂

juv.

Smaller than Greenfinch, with a longer, paler, more pointed bill than Serin. Adult male is colourful: yellow face and breast, bright rump, strong yellow wing-bars and tail-sides. Black crown and cheek, green back and streaked white belly. Female duller, with streaked back and underparts but similar wing pattern. Moves in restless flocks, often with other finches. Call a sharp, rising *tswee*.

Where to see Common breeder throughout Finland. Migratory in the north, but many winter in the south. Found in forests, parks and gardens.

Snow Bunting *Plectrophenax nivalis* 17cm

♂

♂ sum.

Unmistakable large bunting. Adult males in summer are mostly white, with black back, wings and central tail feathers. Females and winter-plumaged males are mainly white with pale brown or grey on back, head and flanks. Bill short, black in summer, pale brown in winter. In flight, it appears very white, showing bold wing pattern. Often forms restless flocks. Call a clear, rolling *chew* or *tju*, frequently heard in flight.

Where to see Uncommon breeder on tundra and high fells of Lapland. Migrates in large flocks in spring; smaller numbers pass through or overwinter in southern Finland and along coasts.

Lapland Bunting *Calcarius lapponicus* 15cm

A little larger than Reed Bunting. Adult male in summer is handsome, with black head extending onto breast. White eyebrow continues to flank, forming a clear border between black breast and reddish-brown nape. Back black with brown streaks; belly white. Female and male in winter are paler, with black spots on breast and reddish-brown tones on nape and head. Juvenile overall light brown and streaked, but shows two wing-bars with reddish-brown between, resembling winter male.

♂

♀

Where to see Uncommon breeder in Lapland, mainly in bushy fell meadows and open tundra. Seen across Finland during migration.

Reed Bunting *Emberiza schoeniclus* 14cm

Adult male in summer has an obvious black head and large chin patch, with a bold white collar and moustachial stripes. Back brown and streaked, rump grey and belly pale brownish-white. Female more brownish overall, with streaked crown, pale eye-stripe, and less distinct collar; chin and breast lightly marked. Non-breeding plumage in both sexes is duller, head pattern partly obscured. Juveniles similar to females, heavily streaked. Call a sharp *tsic*. Song a short, simple series of repeated notes.

Where to see Common breeder throughout Finland, especially in wetlands, reedbeds and damp meadows.

Little Bunting *Emberiza pusilla* 13cm

Smaller than other buntings. Bill is thick, fairly long and dark. Head with chestnut-brown cap, cheek and chin. Distinct dark and pale stripes with clear pale eye-stripe. Back brown, streaked darker; underparts whitish with fine streaks on breast and flanks. Sexes alike, female and juveniles slightly duller and less contrasting. Call a sharp, dry *tik!*, softer than Rustic Bunting. Song a thin, high-pitched trill, not loud.

sum.

Where to see Uncommon breeder in the northeast, rarer than Rustic. Eastern migrant, found in small numbers during passage, often in bushes, birches and thickets.

Corn Bunting *Emberiza calandra* 18cm

Larger and bulkier than other buntings, with a strong head and thick bill. Plumage is plain brown, lark-like, with heavy streaking on the upperparts and paler underparts with fine streaking across the breast. Overall, it looks rather plain, lacking bright colours or striking markings.

Where to see A rare visitor to Finland, recorded only a few times each year. Found in open farmland, fields or wasteland, where it may perch on trees, bushes, poles or fences.

Rustic Bunting *Emberiza rustica* 14cm

Bill is thick and fairly long. Adult male in breeding plumage has black head with white eye- and cheek-stripes. Breast, back and upper back reddish-brown; throat and belly white. Female similar, but head brown with less contrast. Juveniles streaked brown and buff, lacking a strong head pattern. Call like Little Bunting, a sharp metallic *tik!* Song is melodious, clear and pleasant.

Where to see Uncommon breeder in the east, mainly in bushes at swamp edges. Eastern migratory bird, seen in small numbers across Finland during migration.

♂

♀

Ortolan Bunting *Emberiza hortulana* 16cm

Similar in size to Yellowhammer, with a striped brown back typical of buntings. Summer males show a greenish-grey head, pale yellow throat and orange-brown underparts. Females and winter males are duller, with heavier streaking, but all plumages display a yellow eye-ring. Song is a soft, melodic series of repeated notes, reminiscent of the Yellowhammer but shorter.

Where to see An uncommon breeder from southern Finland north to Oulu. Prefers open fields and ditches with trees or bushes for song posts.

Yellowhammer *Emberiza citrinella* 16cm

A medium-sized, thick-billed bird with a striped bunting-like appearance. The adult male is striking, with a bright yellow head and underparts, and a reddish-brown breast and rump. Females and winter males are duller, greyish-brown overall, but always show at least some yellow in the plumage. Song is a familiar, easily recognised phrase, *si-si-si-si-si-suue*, often delivered from treetops or bushes.

♀

Where to see A common breeder, especially in southern Finland, becoming scarcer farther north and absent in northern Lapland. Found in trees and bushes at field edges and frequently in logging areas with scattered vegetation.

FURTHER READING AND RESOURCES

BOOKS

Svensson, L., Mullarney, K. and Zetterström, D. 2022. *Collins Bird Guide*. 3rd ed. HarperCollins, London.

WEBSITES

The Bird Atlas of Finland, coordinated by **BirdLife Finland** and the **Finnish Museum of Natural History**, provides comprehensive data on the country's breeding birds through the work of thousands of volunteer observers. The latest survey of Finland's breeding bird populations was conducted during **2022–2025**.
lintuatlas.fi/bird-atlas/

BirdLife Finland is the leading organisation dedicated to the conservation of birds and their habitats in Finland. It is part of the global BirdLife International network.
birdlife.fi/in-english/

Birdphoto.fi is a specialised website which showcases high-quality photographs and sound recordings of European and Finnish birds. It serves as a resource for editors, publishers, and nature enthusiasts, offering tens of thousands of bird images across hundreds of species, along with audio clips of bird calls.
birdphoto.fi

NatureGate is an online platform that helps identify Finland's birds, plants, and other wildlife, offering photo guides and information to support nature study and citizen science.
luontoportti.com/en

Tarsiger.com is a Finnish birding website providing high-quality bird photographs, recordings and observations. It is maintained by a group of dedicated birdwatchers and photographers sharing their passion for birds worldwide.
tarsiger.com/?lng=en

SMARTPHONE APPS

Collins Bird Guide
eBird
NaturaList
Naturegate

PHOTO CREDITS

All the photographs in this book were taken by **Markus Varesvuo** with the exception of the following images. Bloomsbury Publishing Plc and the authors would like to thank the following photographers for providing copyright material for this book:

Dick Forsman 122 Lesser Spotted Eagle, **Daniele Occhiato** 67 Little Crake male, **Mikko Oivukka** 3 Red-flanked Bluetail, 22 Siberian Jay, 113 Tengmalm's Owl juvenile & 205 Bullfinch juvenile, **Jorma Tenovuo** 67 Spotted Crake.

ACKNOWLEDGEMENTS

Jim Martin and Amy Hodkin (Bloomsbury) kindly commissioned and edited this book. Marianne Taylor and James Lowen, authors of *Birds of Italy* and *Birds of Spain*, respectively, helped review the scope and style. Mikko thanks Markus Varesvuo for inviting him to create this book, as well as his family, loved ones and the entire Kuusamo Nature Photography team for their patience. Markus thanks Daniele Occhiato, Dick Forsman and Jorma Tenovuo for their generous photographic contributions.

INDEX